Scient

Scientifical Americans

The Culture of Amateur Paranormal Researchers

SHARON A. HILL

McFarland & Company, Inc., Publishers
Jefferson, North Carolina

LIBRARY OF CONGRESS CATALOGUING-IN-PUBLICATION DATA

Names: Hill, Sharon A., 1970– author.
Title: Scientific Americans : the culture of amateur paranormal researchers / Sharon A. Hill.
Description: Jefferson, North Carolina : McFarland & Company, Inc., Publishers, 2017. | Includes bibliographical references and index.
Identifiers: LCCN 2017046997 | ISBN 9781476672472 (softcover : acid free paper) ∞
Subjects: LCSH: Parapsychology—United States.
Classification: LCC BF1028.5.U6 H55 2017 | DDC 130.973—dc23
LC record available at https://lccn.loc.gov/2017046997

BRITISH LIBRARY CATALOGUING DATA ARE AVAILABLE

ISBN (print) 978-1-4766-7247-2
ISBN (ebook) 978-1-4766-3082-3

© 2017 Sharon A. Hill. All rights reserved

No part of this book may be reproduced or transmitted in any form or by any means, electronic or mechanical, including photocopying or recording, or by any information storage and retrieval system, without permission in writing from the publisher.

Front cover images © 2017 iStock

Printed in the United States of America

McFarland & Company, Inc., Publishers
Box 611, Jefferson, North Carolina 28640
www.mcfarlandpub.com

Table of Contents

Preface   1

Introduction: Popular Paranormality vs. Skepticism   5

1. Amateur Research and Investigation Groups (ARIGs)   13
2. The Paranormal in Popular Culture   27
3. Ghost Hunters and Paranormal Investigators   45
4. Seeking Monsters: Bigfoot and Other Cryptids   56
5. UFO Spotters   69
6. Twenty-first Century ARIGs   81
7. Science and the Public   100
8. Science and the Paranormal   119
9. ARIG Portrayal of Science to the Public   132
10. Methods and Evidence   156
11. Inquiry and Investigation   180
12. Pseudoscience   187

Conclusion: Beyond the Veil   201

Appendix: Ghost Hunting Guidebooks   221

Chapter Notes   231

References   235

Index   241

Acknowledgments

I want to express deepest thanks to those who contributed in inexpressible ways to the completion of this project, with special acknowledgment of the following: my family; personal supporters and patrons of DoubtfulNews.com; Kenny Biddle for being "Awesome"; Ron Bolton for the invitation to Fort Mifflin; Jeb Card for countless suggestions, information, and advice; Jason Korbus and Bobby Nelson; Sharon and Matt Madison; Chip and Grace Denman for use of a much-needed quiet retreat; Torkel Ødegärd; Ben Radford; and Jim Veihdeffer for proofreading and editing suggestions.

Preface

As a skeptic who has not found compelling evidence for the paranormal, yet remains fascinated by these subjects, I'm often asked how I ended up researching and writing about strange claims and the people who investigate them. The short answer is that I love the idea of ghosts, Bigfoot, and UFOs. I expect I always will.

My childhood was rife with media on paranormal subjects. The first books I chose for myself were on haunted houses and monsters. I was an avid viewer of the real-life mini-documentary TV show *In Search Of...* hosted by Leonard Nimoy. I accepted that Bigfoot and the Loch Ness Monster were out there waiting for us to find. I read and wrote ghost stories hoping I would get to experience this phenomenon myself someday. Alien visitation viscerally terrified me and demon possession was about the scariest thing I could imagine. In later years, influenced by scientific training and skeptical literature on these topics, I lost my confidence that these claims were as witnesses concluded. Eventually, I realized that they were still interesting but in a new way. Reports of paranormal phenomena are worthy of serious research if only because they are so ubiquitous and influential in human experience. Whether the cause is in our minds or out of this world, paranormal experiences profoundly affect the people who have them. The event can change a life for better or worse.

I can't recall when I noticed the glossing of paranormal discussion with sciencey-sounding concepts. Such language had always been there. Realization that symbols of science were a key part of portraying paranormal investigation and that non-scientists were claiming a scientific role dawned on me gradually as I became personally and professionally involved in the public understanding of science. Feel free to categorize me as a science "cheerleader" since I am not quiet about recognizing the great accomplishments derived from scientific research. Yet, I understand that science is done by people who aren't flawless or unbiased. Scientists are not better people than non-scientists, but as with any other profession, practicing scientists acquire different skills and experience beyond that of most amateurs. Science is a discipline, not a hobby. Even though science is not perfect, it still is superbly useful and there is no justification for non-experts to usurp earned authority of professionals.

Admittedly, the public does not have easy access to the insider beat of science or the depth of technical knowledge to understand the foundation of scientific consensus on a subject and make a fully-informed judgment on many technical topics. Instead, they rely on media's "balanced" presentation of the story which, unfortunately, often includes equal space for enthusiastic but mistaken supporters of a scientifically untenable, wrongheaded idea. Many people are impressed with the seemingly sophisticated language and imagery used by non-scientific amateurs, but attempts to fake scientific credentials are often blatantly obvious to anyone who is familiar with science. They are not being scientific in their explanations, they are being *scientifical*—a word I use in this volume to connote the attempt to be scientific without achieving it. This book confronts sham inquiry and pseudoscience of paranormal protocols and concepts as followed by many modern researchers.

Individual and social belief and practice regarding the supernatural, paranormal, and the mystical is found in every society and culture. The term *paranormal*, so popular in American entertainment in the early twenty-first century, evokes the discussion in this volume of the complex and nuanced view American society has about science—as an authority, an icon, an image, a mindset, a community, a mission, a gimmick, or a scapegoat. Uncritical acceptance of paranormal explanations may ultimately lead to invoking other mistaken beliefs or result in faulty conclusions, it can cause unhealthy fear and distress, and it may distract from responsibilities of real life and real problems to be solved. Ignoring this important topic would be, paradoxically, unscientific.

Repeated surveys consistently demonstrate that paranormal belief is common among at least half of the U.S. population, that this has remained so over many decades, that it is a reoccurring and important theme in popular culture, and that it is part of the human experience to entertain thoughts regarding that which may be beyond our mundane daily environment. Per the Baylor Religion Surveys of 2005 and 2007, more than half of Americans say they have researched at least one paranormal topic. The alleged prophecies of Nostradamus, for example, is an extremely popular subject as many people find the idea of a historical psychic seer who advised great leaders of his time as romantic and mysterious. Ghosts, UFOs, and cryptids (hidden animals) are close behind (Bader et al. 2010). We've all read some book or article or watched a documentary about these topics at some time in our lives and wondered how much of it was true. If you believe the extraordinary claims are or could be true, you are far more likely to seek out more information about them. In *Paranormal America*, Bader, Mencken, and Baker (2010) see indications that paranormal belief will likely continue to grow in the United States. View the expanded cable selection on television on any night and odds are you will find a program depicting paranormal investigation.

Media coverage and television popularity alone suggests that paranormal investigation is trendy. There are many possible reasons why this is so. No doubt some get a rush at the thought of facing scary situations. People frequently participate in paranormal investigation activities to satisfy personal curiosity about an unsettling experience they have had themselves. Socializing with like-minded people who appreciate the same esoteric subjects is enjoyable. Or, a person may find an ego boost by being part of something that science "can't" solve. Participation in an investigation like those depicted on TV empowers an individual. Viewers see it being done on TV and think, "Hey, I can do better than that" or "That looks like fun." Involvement in fringe activities could be as simple as escaping the drudgery of daily life and finding novelty in the mundane (Sagan 1972). In a more philosophical sense, Molle and Bader (2013) suggest that the lack of control and involvement people have with modern technology and complicated science results in the "embedding" of these difficult concepts into structures they feel they can control and participate in, such as ghost hunting.

The core of this volume is my study of amateur research and investigation groups (ARIGs) who are passionate about extraordinary claims and the paranormal. In order to fully understand why it was useful to examine this section of American society, I provide further justification for why paranormal themes are worth paying attention to. I also describe how these groups gained popularity. Finally, I explain what function they serve, what outcomes they produce, and how they reflect science in American society.

Many sources cited in this volume go into depth on various aspects of science, the paranormal, psychology, sociology, and the media. There hasn't been a work that focused on how (and perhaps why) amateurs use science in a portrayal of paranormality. My findings reflect not only on the media influence of active paranormal investigation in America but the state of science appreciation and education as well. The passionate pursuit of the mysterious and unknown speaks volumes about meaning we seek in life (and death) and how we view the world. Questionable conclusions and claims about the paranormal are presented to the public as authoritative. Therefore, I have inserted myself into the conversation about popular paranormality to challenge those conclusions and even the claims themselves.

Yet, we must be clear that people who have paranormal experiences and interests are not weird, crazy, or stupid. And, scientists are not cold, unfeeling, and closed-minded. To make progress in understanding, those limiting stereotypes must be discarded. We should be able to discuss these issues through civil discourse, not insults and *ad hominem* attacks.

Researchers and readers interested in a mysterious universe, hopefully, will not see this book as mean-spirited but as constructive criticism that leads forward toward progress. The experience will be a bit like looking in the mirror, I'm afraid, and it will not always be a flattering reflection. Please don't take it personally. The patterns I identify should be changed but I'm not discouraging research into fringe fields. I am, instead, encouraging new and better approaches to paranormal field research and investigation—jettisoning the wastes of time and effort, rejecting the media-derived expectations and preconceptions, and reflecting upon the ultimate point of this effort. An amateur researcher who can thoughtfully explain what he intends to do and exactly why has a worthwhile mission—more power to him. I discovered that in digging deeper, past what the popular media presents as paranormal investigation, the discoveries become more illuminating. See if you think so too.

Introduction: Popular Paranormality vs. Skepticism

Fort Mifflin

It was a warm, clear summer evening at Fort Mifflin, a decommissioned military site and National Historic Landmark, located on the Delaware River just outside of Philadelphia, Pennsylvania. An online acquaintance graciously invited me to join his paranormal investigation team and two other such groups to experience the Fort's ghostly reputation. Some locations harbor a strong "sense of place." Historical sites like Fort Mifflin feel like sacred ground, imbued with meaning and holding wisps of memories of past events. Access to the Fort after public hours is provided by appointment to groups who seek evidence of the paranormal. I was along as a guest to observe the researchers and to see if I could experience at least some of the phenomena they assured me was common at the location. I felt privileged to walk along the Fort walls at sunset and to silently appreciate the efforts undertaken here at peace and during battle. Then, I returned to the well-lit hall to watch the teams unpack their cases of equipment and gear up for the night ahead, stashing extra batteries, flashlights, cameras and gadgets in their pockets.

Small groups of three to five people fanned out to various areas to try their luck at communicating with spirits of Revolutionary and Civil War officers, soldiers, and prisoners who passed through this walled fortress in the swamp. When the groups came back together to get fresh batteries for their gear and reapply bug repellent, they reported what they found. Those in my group placed their electromagnetic field meters (EMF) on a ledge or table inside the stone buildings and asked questions into the darkness. The constant drip of water, chorus of frogs, and droning mosquitoes around my head were loud in the otherwise peaceful night, except for the occasional deafening cargo plane coming in for a landing at the adjacent international airport. Green and red lights of EMF meters flashed. Seasoned participants claimed to feel a presence beside them, even lightly touched as if by a spectral breeze, as we stood elbow to elbow listening intently in the stifling stone structures. As the night progressed, the groups reported experiences. One lucky guy from another team said he saw a full-body apparition in Revolutionary period clothing. Another participant reported seeing a shadow figure run outside the entrance gate. As the participants stood around chatting, they discussed their current cases back home, one of which involved a child's haunted room, and another, an exorcism they attempted.

I was not touched by any spirits this night, only by the hordes of vampiric insects. Yet, the human behavior I observed was curious. More irritating than the biting insects was the unimpressive evidence uncritically claimed to be "paranormal" that suitably impressed everyone else. I saw hype and hope but nothing convincing about these famously haunted grounds. Paranormal enthusiasts obviously love these experiences. They buy gadgets and spend their weekends in old buildings listening for the tiniest taps or thumps, ultimately concluding they have had a paranormal encounter.

I went back to my hotel room well after midnight and ruminated. There was something to this experience, but I did not grasp it. Amateur investigators seeking ghosts were finding something else while searching for the paranormal—hope, purpose, and excitement. They were, to my eyes, playing scientist in an attempt to order and reinforce their view of the world.

Missing was a curiosity about what might really be going on, a drive to dig deep into other explanations, and any inclination to make non-paranormal conclusions. Too many of the amateur investigators I studied for several years sought a paranormal explanation as a goal, and then gave up when they felt they achieved that to their own satisfaction. Questions asked of the spirits into empty air didn't seem to be the right questions. I heard no sufficiently satisfying replies.

There is a difference between enjoying a subject or activity as a hobby or entertainment and publicly claiming expertise and authority in that field. Those interested in paranormal and unusual claims cross that line by delivering information and opinions to the public. As a public declaration, this means criticism of it is fair game. This I have done for years. This book examines how amateur researchers present themselves to their clients and the public as serious and scientific investigators of the paranormal—a topic not explored in depth before. My intent is not to promote or discourage belief but to consider what ghost researchers, ufologists, and cryptozoologists present to the public considering the history of these fields, the media

Fort Mifflin in Philadelphia, Pennsylvania, welcomes paranormal investigators for after-hours visits. Investigators report multiple sightings of apparitions of soldiers, collection of EVPs, and appearance of strange lights. Photograph by Kenneth Biddle.

influence, and the standards of scientific investigation. Theirs is a narrow view of these topics—they follow what they have seen others do. The ideas to which they subscribe are deeply enmeshed with their personal identity.

The approach I take is broad and is that of thoughtful skepticism. I question where these researchers obtained their strongly held opinions about the paranormal, investigation, and science. By the end of this survey, I will volunteer my suggestions to improve approaches to investigation of these strange experiences and claims to better illuminate what is physically there. Or not.

No Nonsense, Please

> *To be skeptical means to harbor reasonable reservations about certain claims.... It means to want more evidence before making up one's mind*—Pigliucci 2010: 137

All of us are skeptical about something we're not prepared to accept. I am skeptical by nature and have intentionally nurtured that conservative habit because I want the best answer, the likely truth. I have cultivated an intolerance to nonsense, but I understand how alluring nonsense beliefs can be.

There are those of us that hold the position that seeking the best evidence, and using reason, logic, and tools of science, is the most effective and reliable way to make truth claims. We eschew the subjective and do not accept weak evidence for an extraordinary claim. These characteristics are the basis for a modern skeptical approach. People tend to throw the concept of skepticism around selectively and shallowly, or they equate it to cynicism (as a bad thing) or to being a debunker who ruins the mystery. You will often hear "I'm a skeptic" or "I'm skeptical" from people who are not sure about or who doubt some claim. That is a common, casual use of the term. Simply calling oneself a "skeptic" is not the same as thoughtfully practicing it, however. When I refer to *skeptic* from this point forward, I mean those who follow a skeptical method to assessing questionable claims. As I wrote in *The Media Guide to Skepticism*, the modern version of skepticism strives to apply a few tenets and objectives.

Respect for the evidence. The application of reason to a body of solid evidence is the best method we have to obtain reliable knowledge.

Respect for methods, conclusions and the consensus of science. Science is a specific way of obtaining information that is designed to reduce the chances of coming to an incorrect conclusion. Using a scientific process will minimize errors (but not eliminate them entirely). So those following a skeptical approach are often vigorous advocates of science—in medicine, in schools, and for informing policy decisions. Fake, junk, and pseudoscience is called out as a ruse. Logic and math are also components of science that can be valuable in assessing claims.

Preference for natural, not supernatural, explanations. Material laws based on careful and repeatable observation of the natural world provide rational boundaries to determine explanations. No external agent or manipulator is invoked. A skeptic will seek an explanation that does not call for a supernatural, unproven (and possibly unprovable) entity to be included.

Promotion of reason and critical thinking. Many skeptics are practiced in identifying mistakes in arguments and reasoning.

Awareness of how we are fooled. People routinely fool themselves and are fooled by others. All of us can be fooled; there is no shame in admitting that we tend to trust what others tell us. Frequently, we over-rely on our senses and memory to provide facts—for example, "I know what I saw" or "I remember it like it was yesterday." Skeptics are wary of eyewitness testimony because observation is fallible and memory is malleable. Stories of events, even from trustworthy people, make for poor evidence on their own. Even collectively, anecdotes can't tell us much about the validity of the claim. Skeptics recognize that people tend to look for, remember, and favor the evidence that supports their preferred conclusion. The human brain is prone to seeking patterns and meaning that may not be genuine. We have a tendency to associate a number of events together in meaning if they occurred associated in time.

Skepticism, like science, is a collective activity. One person can't go it alone. To improve critical thinking requires learning, practice, discipline, and seeking other sources for input. It's work. It's easy to doubt things just because it feels like you should; proper skepticism involves understanding why one should accept or deny a claim (or to withhold judgment) and to elucidate the reasons for doing so.

I've not had an experience I consider paranormal but exploring the topics has changed me. As hard as it may be for me to accept an eyewitness' conclusion that they experienced the paranormal, a person's experience is their own. I will not disparage your experience or try to convince you of my point of view. Psychological research has shown that doing so can create a "backfire effect" that causes people to double-down to defend their own position.[1] Therefore, my approach will be to present my view and you can freely decide what to do with it.

After years of studying the paranormal, including a significant body of academic and skeptical sources as well as primary accounts, I have discovered that the array of explanations for any paranormal experience is possibly limitless. We can't ever exhaust all the options and we may not be able to obtain the necessary information to conclude a reasonable explanation for all claims. I'm OK with that. What I'm not OK with is those who jump too quickly to the paranormal conclusions and promote these to the public. My deep interest in paranormal themes in popular culture in America is not dependent on belief, it is informed by evidence and guided by practical skepticism.

Worldviews

In the excellent book *Worldviews: An Introduction to the History and Philosophy of Science*, Dewitt (2004) explains how we construct our personal vision of the way the world works. Just like a mosaic or puzzle, everyone connects their own pieces together from experiences, learning, feelings, beliefs, and hopes. This is our "worldview." Beliefs can be acquired from those around us—taught to us in childhood, or absorbed through our cultural products or social interactions—or from our conclusions based on life experiences. We try to fit our subsequent experiences and beliefs into this puzzle. From within a particular worldview, whether based on scientific consensus or supernatural beliefs, the facts and explanations seem to be common sense, correct, and obvious. It is difficult to step outside of that sphere and consider a different worldview. Most people won't. Therefore, it's uncommon to ditch the puzzle entirely and start blank. But the puzzle pieces can be modified or gradually replaced by other pieces if we allow it. Eventually, the entire picture can change and a new worldview is in place. A shift in worldview is often a slow process but can be dramatic and life-changing. If you accept your life is normal one day but the next day you are convinced everyone is playing a role in a virtual reality scenario, for example, everything looks and feels a lot different.

Beliefs color our observations leading to bias and perception errors. Beliefs are tied to how we behave and how we identify ourselves. So, unfortunately, we can be easily lulled into only acknowledging the evidence we favor to support our beliefs, our worldview, discarding the rest. We will interpret facts in a way favorable to our beliefs, or we may accept what we are told from those we see as authorities. Those who have a strong, often emotional investment in paranormal beliefs are reluctant to discard those cherished beliefs and will berate and reject the skeptical approach in examining them (Hess 1993). I understand. No one wants the skeptic to expose the flaws in beliefs we hold dear.

Our worldviews allow us to function and make meaning out of the world every day. Paranormal beliefs don't color my worldview anymore, although they used to. I can rather easily slip back to imagining what it's like to hold a paranormal worldview and "put myself in someone else's shoes," at least for a while. At this point, however, I don't just want to believe; I want to *know*. I want to scientifically know, not just "know" in my gut or bones. I want knowledge based on reliable information that would be convincing to others as well. This degree of knowing takes a good deal of effort to achieve. It's not always fun, either. At Fort Mifflin, as the skeptic, I was the unwanted outsider. There was no mistaking that impression. But it was important that I challenge my assumptions, which I did.

An extremely common reason given for accepting a paranormal reality is that so many people report these experiences and believe in them that there must be something to it. This argument from popularity is called the "bandwagon appeal" (Pigliucci 2010); we all want to jump on. The skeptic is not always keen to jump on the bandwagon, though, and is not afraid to tell the other riders why she refuses. We need skepticism to sift through the deluge of information to obtain reliable knowledge from big scientific questions to personal decisions about finances and health. I accept an awful lot, but I doubt a lot more. It's OK to hold a provisional conclusion while waiting for better information to come along. And it is OK to just say "I don't know" until you actually do know and can back it up.

Organizing Principles

Some category of paranormal belief is widespread among at least half of the U.S. population, and this has remained so over many decades, indicating that it is part of the human experience to entertain thoughts regarding that which may be beyond our mundane daily environment. Amateur researchers spend considerable time exploring these topics. The serious researchers directed their aim toward exploration of specific claims and locations. They carefully structure their activities to attempt to understand or conclude facts about these topics and to satisfy their own interest. It is a human endeavor with which I fully sympathize and encourage.

My effort to examine and reconcile the difference between amateurs' work and the scientific community is explained thoughtfully by Dolby (1975) who wrote about how a scientist should approach a differing viewpoint than one's own:

> A scientist ... should try to become familiar with the conceptual and methodological approach of his opponents. He should try to see what is appealing about the other view, what is accepted as fact and what methods of reasoning are most acceptable.
>
> He can enter into a dialogue with his opponents. He should try to put his critical reaction into forms which will be acceptable to his opponents....

I do not categorize the amateur investigator as an "opponent," but they do have a viewpoint very different than mine. I could not fairly comment, criticize, or reject their efforts without attempting to understand what, how, and why they do what they do. What I concluded is deeper and more nuanced than skeptics and non-paranormalists assume—the people and their goals are worthwhile; it's the structure they work within that is flawed. That structure has been erected and encouraged by our popular culture and is now entrenched.

Ghost hunters, Bigfoot trackers, and UFO chasers pursue the unknown; it's exciting, dramatic, romantic. Consulted by media as "experts" in the paranormal and even referred to as parapsychologists (a PhD-level title), they may be called into difficult social situations where trained experts would be recommended. (See examples in Houran & Lange 2001.) The public sees them as educated specialists, maybe even brave explorers or minor superheroes. Some have become celebrities and in demand at events. They are seen as "scientific," serious professionals. I try to explain why social cachet occurs in various threads throughout this book. As we see with climate change, and previously with the concept of evolution of life on earth, the public overall is not adept at understanding the depth and breadth of scientific consensus on a subject. Are there ghosts, UFOs and monsters to find? Or is this a snipe hunt?[2] I surely don't know all of what might be out there in the dark, but I do know that the belief that there is something extraordinary to find is a powerful impetus for humans to go out and look. They spend huge amounts of time, money, and emotional and physical effort doing it.

Much of the information and attitudes of what I term *amateur research and investigation groups* (ARIGs) contained in this volume was gathered in 2010; therefore, many of the quotes will no longer be available on the web due to the ephemeral nature of websites. Also, some ARIGs may have changed their views and would not relate the same sentiments today. But, the general concepts, the purveyance of paranormal ideas, and the methods of investigation remain basically unchanged. The people may have moved on, but the cultural niche and associated behaviors remains. It's time to think more deeply about why and how to research fringe topics and if answers will be found via scientific methods or what people *think* are scientific methods.

1

Amateur Research and Investigation Groups (ARIGs)

The world is full of curious, passionate amateur inquirers. This book is very much about the average American who has engaged in activity about which he or she has strong feelings and concerns. *Amateur* literally means "one who loves." Those who research paranormal topics as hobbyists fit this description well.

Labeling people as "amateurs" can be a way of discrediting their contributions (Lyons 2009). However, I am not using the word in this disparaging way. An amateur's passionate dedication is admirable and inspiring. We are all amateurs at something—fascinated with a subject to which we commit considerable time and money to gain understanding. Though amateurs are motivated out of love for the subject, with that comes a normal human desire to be acknowledged for their contributions, as they should be (Mims 1999). But it's not easy for the academic outsider to be recognized. There is a no man's land in the area between amateur and professional scientist. Disputes regularly arise.

Science as a serious vocation emerged by the 1870s and become established so that insiders in each field gained an air of eliteness based on their specialized knowledge. Progress in gaining new knowledge in the sciences quickly outpaced the public's ability to obtain and understand it. As this happened, science became a professional arena. But this did not mean amateur contributions disappeared. Amateurs are free to explore any subject they wish, to be innovative, unencumbered by the seemingly endless task of finding funding, producing journal papers, teaching, and advancing their own careers. They are typically working on topics and concerns abandoned or ignored by institutionally affiliated scientists. Paranormal researchers have taken advantage of these seeming voids. Yet, these topics only *appear* to be abandoned. As discussed later, scientists did seriously examine these fringe topics, decided there was nothing there to pursue, and then moved on. Alternatively, these subject areas became valued in cultural or psychological studies.

The American public remains fascinated with possibilities of paranormal reality. It was rare to see official scientific exploration into field studies of ghosts, UFOs, and strange creature sightings. With reported cases growing and gaining media attention, the amateurs stepped in, fueled with interest and enthusiasm to meet a personal and public need. Amateurs spend their own money to travel and investigate the subjects they are passionate about. They buy books and equipment. They will also mirror social and academic societies by maintaining websites and organizing meetings and conferences. Despite the paranormalists who have found a measure of celebrity through television, books, and the Internet, very few will financially profit from these activities. They will likely not land a paying job on television (or even a paid interview) and related business ventures such as ghost tours, paid investigations, educational offerings and book publishing, or expeditions to find the quarry won't be financially successful. Thus, money is a secondary concern for many who pursue this passion as it appears to provide great personal enjoyment and meaning in their lives.

Leisure Time

Americans love their leisure time. Indeed, this time is so critical to our lives that we schedule our leisure activities—like watching TV shows, sports, seeing movies, reading, even volunteering—as things we *must* do. These activities become a necessary part of life. How we spend our leisure time plays a key role in defining who we are. What we do outside of our job or family responsibilities provides an opportunity to expand our self-identity, to provide an outlet for personal expression, and explore unfulfilled potential. Our "real" jobs may be viewed solely as the means to sustain leisure activities and interests. Many of us socially define ourselves more by our leisure activities than by our careers—the animal lover, sports fan, history buff, auto enthusiast, etc.

The concept of "serious leisure" was proposed by sociologist Robert A. Stebbins in 1982. Stebbins defines this as "the systematic pursuit of an amateur hobbyist or volunteer activity that people find substantial, interesting and fulfilling." This activity allows for expression of special skills, knowledge, and experience (Stebbins 2007). Our practice of these activities gives us great personal and social rewards and enhances our lives. We feel better about ourselves, like we have a purpose. It is rewarding (Stebbins 1982 and 1992). For some, pursuit of paranormal research and investigation is serious leisure, incorporated as part of their lives, and integral to the way they view themselves and present themselves to others.

Defining Amateur Research and Investigation Groups

I am interested in how amateur research and investigation groups work, behave, and utilize the concepts of science and skepticism. The group, not individuals, is my unit of study for this project, although much of the conclusions can be applied to individuals as well. Not all these groups would self-identify as "paranormal researchers" as some find the term "paranormal" limiting or suggests less-than-serious goals and methods. Personal belief was of less interest to me than the methods and framework they used for investigating strange claims which left open the inclusion of groups who took a skeptical approach. Therefore, I coined a more inclusive term—Amateur Research and Investigation Group, or ARIG (pronounced AIR-rig). Criteria that define the amateur research and investigation group (ARIG) are as follows:

1. *Not under the auspices of an academic institution or headed by working scientists.* This excludes organizations such as the Institute of Noetic Sciences, the Parapsychological Association, or a university-affiliated lab. Paranormal investigation groups headed by a working scientist as part of academic research may be common in fictional depictions but are rare to non-existent in real life. Such groups would also be outside the scope of this research as they would be significantly different than the typical groups in operation around the country.
2. *Activities pursued focus primarily on unexplained events such as reports of hauntings, mystery animals, unidentified aerial objects, natural anomalies, and parapsychological phenomena.*
3. *Having a grassroots process of formation and organization, but may hold some affiliation with a larger group.* One characteristic of these groups is that they can be independent to a fault, dissolving and reforming repeatedly. They are almost exclusively self-sustained by their members' contributions, donations, or fundraisers.
4. *Advertisement of the group, and their activities and/or services via the Internet.* Internet presence is a must-have for these groups as a major means to gain new clients, cases, media attention, and

credibility. Even as word-of-mouth between friends, family, and neighbors is the primary means to be introduced, those interested in contacting an ARIG will typically seek out information about them online.
5. *Activities undertaken by such groups do not provide a primary form of income for participants.* Most groups do not charge for services. In some cases they may ask for reimbursement of expenses incurred for travel. Some are registered non-profits or state they are not businesses. Some groups are affiliated with tourism activities, such as guided tours, shops, or museums, or will promote books or other merchandise sales which generates some income that presumably is used to sustain the group's activities. Those few groups that have members with some degree of notoriety will charge public or private appearance fees. But, in general, paranormal investigation is not a money-making venture, but, like most hobbies, is an expense.

Most groups also espouse attitudes and behaviors such as a sense of doing important work, helping people, and contributing to a higher good. Many group members dress alike to enhance group cohesion and unique identification. The stereotypical group uniform is a black polo or t-shirt with the team name and logo, a conformity that appears to be derived directly from the TAPS *Ghost Hunters* television show. I'll have more on TAPS further on in this chapter.

Appeals of participating in an ARIG are the excitement of discovery and the desire for enlightenment. There is a thrill in seeking out these experiences, sharing an adventure with others, and the feedback between group members that reinforces a belief (Bader et al. 2010; Childs & Murray 2010). Many ARIGs openly announce that they will be the ones to provide proof to the world of what they know is out there (Sykes 2016). With minimal training, no advanced degrees, and subjective techniques learned quickly, anyone can become experienced at this activity (Bader et al. 2010). ARIG participants are involved in a real investigation that feels edgy, dangerous, and groundbreaking. It's hands-on, down and dirty, paranormal immersion.

In addition, there are other more personal aspects appealing to ARIG participation such as being social with others of like-mind, being part of a group, and gaining a sense of identity and importance. Many ARIG participants will cite their own experiences with a haunting, Bigfoot, or a UFO as their impetus for involvement. They want to explore it deeper with others who shared a similar experience.

The Experience

Experience is at the core of human existence. As social creatures, we share our experiences with others. Effective stories are extremely convincing and emotional. The act of sharing with others not only spreads information, but strengthens the memory and meaning of the experience for ourselves. In turn, meaning and experiences from our lives color our interpretation and memories of events we have witnessed (Loftus 1996).

Bader et al. (2010) saw that there is great importance placed on firsthand experience and a shared interpretation of events. Personal experience and interpretation will trump contrary or critical commentary from outside experts every time. Those who have seen a UFO, had a strange animal encounter, or thought they had a ghostly experience were far more likely to take up their own research. Any such experience was a powerful indicator that further information on it would be sought, even to the point of joining a group to conduct specific investigations on the subject.

The structure of the experience will shape the reaction. An unnerving environment or participants primed with stories about previous strange events has an improved chance of producing more thrills and chills. Group dynamics, including the reactions of others, will influence and reinforce the experience ("Feel a chill?" "Yeah, I felt that too!") (Hall 1972; A. Hill 2010; McRae 2012). Ghost and monster stories are ancient. The quest to seek out these unusual experiences is what Annette Hill calls the "culture of reenchantment."[1] A paranormal experience can evoke confusion and intense emotion. In my exploration of these topics, it's common for an ARIG member to say that an experience they had was "life changing." They seek participation in an ARIG as a way of working through that belief (A. Hill 2013). Folklorists use the word *ostension* to describe the behavior of living out a legend. The legends—haunted houses, Bigfoot, visiting alien craft—are woven into a narrative for ARIGs to seriously pursue with the aim of having a powerful experience.

Citizen Science

Amateur naturalists are passionate about observing and collecting specimens in their leisure time. These citizen scientists contribute to new knowledge by sharing their finds with professionals who aggregate and verify the data sets (Regal 2011). In areas where considerable and prolonged direct observation and collection is required (such as astronomy, meteorology, fossil discovery, and animal population studies like bird counts), amateurs still contribute valuable data and knowledge (Gordon 2015; Gregory & Miller 2000; Lankford 1981; Lyons 2009; Mims 1999). Scientists still rely on amateurs to contribute data by keeping records and identifying new discoveries for further investigation. Computers have expanded the capabilities of amateur contributors (Mims 1999) by allowing access to tools, resources, and expert advice. There is no doubt that amateurs can be important to advancing scientific knowledge. You don't have to have a university degree or accreditations to participate in the scientific process, just follow a proper procedure and communicate the information reliably. "Citizen science" projects designed and managed by academic or professional scientists are gaining in popularity. ARIGs are different from citizen-science activities. While both ARIGs and citizen science projects are made up of volunteers usually without scientific training who participate in scientific observation, data collection, or processing, I came across no citizen science projects related to the paranormal. Under the auspices of scientific institutions and academic researchers or local professional groups, citizen science projects have clear, measurable goals and strict methodology. The results are subjected to expert review. Examples include bird and butterfly observation logs, animal mortality counts, Moon crater counting, and classifying galaxies based on shape. Citizen science is a form of public outreach to create interest and appreciation as well as tap into amateur enthusiasm. Gordon's (2015) *Sasquatch Seekers Field Manual* was the first explicit mention I had seen of "citizen science" investigation of paranormal claims. He notes that those witnesses who simply claim they know Sasquatch exists because "I seen um!" provide nothing useful to securing information. Nonetheless, Gordon describes these eyewitness accounts as a preliminary stage of gathering data. However, Gordon misses key requirements for amateur science to be helpful—training in observation, collection, and recording, and an overseeing, credentialed scientist willing to examine the submissions. Field manuals for amateurs to collect anomalous observations are not equivalent to obtaining training from professional scientists. Dozens of various "guides" to spotting cryptids, ghosts, and UFOs promise to make you into a real investigator, just like the people you see on TV. The ease of self-publishing and the popularity of paranormal investigation have resulted in a plethora of instruction and guide books that span a quality spectrum. Individual ARIGs write their own manuals to train recruits in the group. Paranormal "How To" books, written by nonprofessionals, are discussed in the Appendix.

A few amateur-led groups have developed databases of sightings and reports of paranormal occurrences. Further on in the book, I discuss the use of these databases and the various explanations (ARIGs often refer to them as "theories") that have developed over time and are used in pro-paranormal circles similar to how scientists would use scientific theories as models.

Organization

In the mid–2000s, amateur ghost, UFO, and cryptozoology researcher groups were recognized but no one had attempted to methodically count them. Only "guesstimates" were available. Word of mouth in the paranormal communities in early 2010 suggested the number of these groups in the U.S. alone had grown into the thousands. A few surveys examined their numbers: Andrews (2007) found 316 ghost investigation groups via a Google web search in January 2007; Brown (2008) found 27 in six New England states, roughly correlated with population. A reasonable estimation was that there were around 2,000 ghost-specific groups as well as many small groups focused on various anomalies. Annette Hill estimated 2,500 ghost hunting groups in 2010 in Britain compared to only 150 ten years prior. The same might be assumed in the U.S. as the explosion of growth was linked to the television popularity of *Ghost Hunters* and similar shows as well as the availability of the Internet, particularly YouTube, as a place to host clips and promotional material.

In the summer of 2010 I methodically examined more than 1,200 websites run by ARIGs across the U.S. Preceding that, I established a pool list of about 1,600 sites that existed on the web by using search terms in various combinations: "investigation," "research," and "group" paired with a descriptor such as "paranormal," "anomalies," "UFO," "ghost," "Bigfoot," "cryptozoology," "scientific," and "skeptical." With a raw set of web addresses (URLs), I eliminated the duplicates, making sure I did not include different branches of the same group, then numbered and randomized the list. Randomization of the starting list was necessary to ensure a representative sample of groups across the country and with various specialties. Typically, indexes of groups are arranged by state so a prospective client can contact one locally. I started down the randomized list, one by one, completing a premade data collection form for efficiency and consistency, until I cataloged 1,000 valid, accessible, qualifying sites. This core body of data was supplemented by additional observations, examples, and quotes to illustrate common themes I observed across ARIGs. The large number of sites examined allowed me to see subsets of the groups that had or lacked certain characteristics. I was interested in a collective result across the spectrum. Individual ARIGs will be different from one another. Generalization can be misleading. But there clearly were broad trends visible across the range of ARIGs.

I documented ARIGs in every U.S. state and the District of Columbia. Several groups had multiple chapters in different areas of the state or in other states. Many groups travelled to neighboring states for investigations. Therefore, there is likely no area that cannot be reached by some ARIG. The eastern half of the United States had a greater number of groups as expected in relation to higher population density. The state with the most groups, 81, was Ohio. Next, was Pennsylvania at 80. Both are certainly underestimates as groups splinter, reform, and go defunct with great regularity. Because of this ephemerality, many of the groups cited by name in this book no longer exist by that name. For that reason, it's not important to focus on individual groups but on their collective behavior. With a moving target of groups in existence, it is impossible to keep track of all active groups at any time. Some groups may not have an Internet presence or consist of one or two individuals operating intermittently.

The Independent Investigations Group (IIG) of Colorado completed a survey of paranormal investigation and research groups (mostly focused on ghosts and hauntings) published in 2012 (Duffy 2012). I compared my results with this survey that located teams in 32 U.S. states, Canada, Italy, the U.K., and Germany. Essentially, not much had changed, but the more in-depth questions of the IIG survey provide additional resolution into the makeup of ghost/ARIGs in particular.

ARIGs are often independent efforts but can be offshoot chapters of a larger group that has expanded or one that identifies as an affiliate of another well-known group or society. Chapters have a direct connection to a headquarters group and operate as an arm of that group. Affiliation is a loose alignment often used just for name recognition. The most common affiliations at the time of my survey was with The Atlantic Paranormal Society (TAPS), the organization depicted on the *Ghost Hunters* TV show. But the other popular affiliate included the Ghost Adventures Crew (GAC), the organization of the *Ghost Adventures* TV show with Zak Bagans. Affiliates are asked to meet certain criteria to maintain connection with the overarching group, yet they operate independently. "Badges" (standard graphic files) were proudly displayed on ARIG websites to indicate affiliations, lending a sense of legitimacy to the group to suggest they are comparable to their more prestigious affiliate. ARIGs affiliated with TAPS report more public recognition since affiliating (Brown 2008). TAPS listed their affiliate ARIGs in the U.S. and abroad on their website.[2] Affiliated TAPS groups were required to maintain certain standards and were expected to adhere to rules, protocols, and ethics[3] so as to be allowed to use the "TAPS family" designation. The criteria for TAPS included maintaining an acceptable web presence.

Affiliation is a useful premise as it creates a common foundation in which to contact other associated groups to share information and cooperate on investigations. It can also serve to standardize methods and behavior; however, if the methods are flawed to begin with, the affiliate family will consist of dozens of copycats going about things the wrong way. Therefore, affiliation is a mixed bag, benefiting the ARIG by associating them with a popular name and a suggesting a sense of validity but creating an army of followers instead of independent innovators. The use of affiliates with a larger national or regional group was rare or nonexistent for ARIGS specializing in cryptozoology, UFOs, psychic claims, or Fortean phenomena though some groups do partner for events or investigations.

With thousands of groups and individuals participating in the research and investigation of claims, progress in their respective field is impossible without the establishment of a foundation of knowledge, guidelines, and standardization of practices. There is little evidence that any standardization exists. There is a modicum of cooperation and community sharing. ARIG results are not consolidated and compared with results from other groups that investigated the same site or incident. ARIGs often do not share their results publicly at all, do not have any method for peer review outside their own group, lack standards or protocols across the field, and have no efficient means to do so. For standardization and sharing to occur, an overarching agency or society acceptable to a large portion of the community would be needed. If researchers followed guidance and standard procedures on use of equipment, processes, and reporting of results, such results could more readily be quantified and analyzed with potentially useful conclusions made. Instead, piles of results sit unchecked and inaccessible, forgotten in filing cabinets, on websites, or on computer drives around the country.

Studies of Amateurs on the Fringe

At the same time as I was formulating my ARIG research, historian Brian Regal was writing the book *Searching For Sasquatch: Crackpots, Eggheads and Cryptozoology* (2011). The book opens with the line "This story tells of dreams that do not come true" (p. 1). In *Searching For Sasquatch*, Regal recounts the colorful history of amateurs and professionals that spent most of their lives invested in establishing Bigfoot as a real creature. They all failed. But for decades, they butted heads and battled for legitimacy and status. Regal's book confirmed many of my own findings regarding "serious leisure" and the attitudes and behaviors of the amateurs and professionals regarding the use of science. Some of the same themes appear in Tea Krulos' *Monster Hunters* (2015). Krulos, a journalist, followed ARIGs and independent researchers into the deep end of ufology, cryptozoology, and ghost hunting. Some of the research in this volume is mentioned in Krulos. He contacted me after reading an earlier article about my explanation of these groups being "scientifical" (discussed in depth in Chapter 9). On his journey with these investigators he had his own share of strange experiences while interacting with those obsessed with lake monsters, Bigfoot, Mothman, demons, nosy aliens, dog-men, and an array of spirits. Krulos captured the personal ups and downs of this work/hobby. Seeking answers about the unexplained, he found, will cause family friction, cost you money and friends, and strain patience in attempts to keep an investigation team together and on track. The disagreements and drama are a side of amateur investigation efforts you won't see unless you are directly involved. Some of the groups Krulos profiled were also in my study and I knew of (or had previously spoken to) several of those included in his book.

Sherrie Lyons' *Species, Serpents, Spirits, and Skulls* (2009) details the transition between amateur and professional science that occurred in Victorian England using examples such as sea serpents, phrenology (study of the mental abilities of a person by examining his or her skull), and spiritualism. Her observations show that knowledge-making shifted from well-respected amateur naturalists to the more rigorous, difficult, and political process of the burgeoning scientific establishment. This was the time of erecting boundaries between professional science and everyone else.

This study is a cross-section of ARIGs in the first ten years of the 21st century. The ARIG website was the *group's* representation to the public. There may have been one person acting as president or leader or web content designer but the website was assumed to represent the team philosophy, procedures, and attitudes. Without the ability to question hundreds of groups individually, I assumed their stated information was what they embodied as part of their interaction with the public. Therefore, this view of ARIGs was very much a snapshot in time within a three-month span of 2010. Yet, in my observations, the conclusions still apply. I took quotes directly from webpages but the sites may no longer exist in the same form. I also provide the caveat that I can't say with any certainty that whomever wrote these statements still subscribes to them. However, I leave current comparison up the reader, and feel certain that sentiments like these still are ubiquitous on ARIGs websites across the world.

Categories

In a broad sense, *paranormal* encompasses extraordinary phenomena perceived to defy current scientific understanding. Within this sphere are subjects of interest to ARIGs such as ghosts, cryptozoology, and UFOs, which have a "spooky" nature, a suggestion of re-enchantment of the world, that makes them popular sources of entertainment.

In this study, ARIGs logically fell into one of four subject categories—*ghosts, cryptozoology, UFOs* and the encompassing *general paranormal*—based on their stated focus. The most popular investigation subject category by far was "ghosts," comprising 879 groups out of 1,000 (87.9%). Additionally, some groups identify themselves as open to investigating "ghosts," "UFO's," or "cryptozoology" and were considered general paranormal. Eighty-one groups (8.1%) stated they were open to looking at cases of all sort of mysterious phenomena without limiting subject areas. However, this value is artificially low since many ARIGs advertising as ghost investigators are open to all paranormal claims and will likely be more flexible if approached with any type of claim. Thirty-five ARIGs (3.5%) identified as exclusively focused on "cryptozoology," the search for mysterious animals, mostly Bigfoot/Sasquatch. But due to the increasing exposure of the field of cryptozoology on television, in monster movies, popular books, via news reports of cryptid sightings, and on the Internet, interest in cryptozoology is expanding. Only two groups were focused solely on UFO phenomena—unidentified flying/aerial objects or anomalous aerial phenomena. However, one of those groups was MUFON, the Mutual UFO Network, a nationwide organization of volunteers that may include several thousand investigators across the country. MUFON is the largest UFO organization in the U.S. with a director in every state (except for the six New England states that are grouped together) and often an assistant director along with a network of investigators. Classification of MUFON as one large group is problematic because it was unique in the data set. Not only was this one ARIG enormous in size compared to others in terms of participating individuals, but the individual state branches had different directors and ways of managing situations in their own state. Attitudes and priorities were likely to be different from chapter to chapter. In total, MUFON had 32 individual chapter websites. Eyewitness reports from anywhere in the nation are sent to local MUFON members for investigation. Because of the overarching organization structure, centralized means of training, and collecting eyewitness reports into a central database, MUFON was counted as one ARIG. Not long after my study, MUFON membership was noted to be in decline and some branches were displeased at the main management of the organization.[4] Some even broke away to be completely independent.[5] But in 2010, MUFON was a large network of UFO investigators, connected and sharing processes and procedures.

The different categories of ARIGs share many features in common, but they have clear distinctions based on the specifics of their chosen focus. It is difficult to generalize each subset of ARIGs, let alone the entire diverse population of participants. They have similarities and some unique differences. And, they evolve through time. For purposes of this book, I mostly discuss all categories of ARIGs but in some cases, specific categories are mentioned and I will use the designation ghost/ARIG or UFO/ARIG, if this clarification is necessary.

A common complaint of ARIGs is that the scientific community does not take them or their interests seriously. This is simply incorrect as is demonstrated through the history of each ARIG field in upcoming sections. ARIG topics at one time were seriously considered by science but lost scientific standing over time. When a field loses scientific respectability for whatever cause, it may be taken up by amateurs who attempt to regain its credibility again. Ironically, some of these fields become the subject of historical or scientific research about religious movements, the nature of science and social, cultural or psychological phenomenon. A consequence of amateurs appropriating a field outside of gatekeeping institutions is the inevitable involvement of exploiters, charlatans, and outright hoaxers.

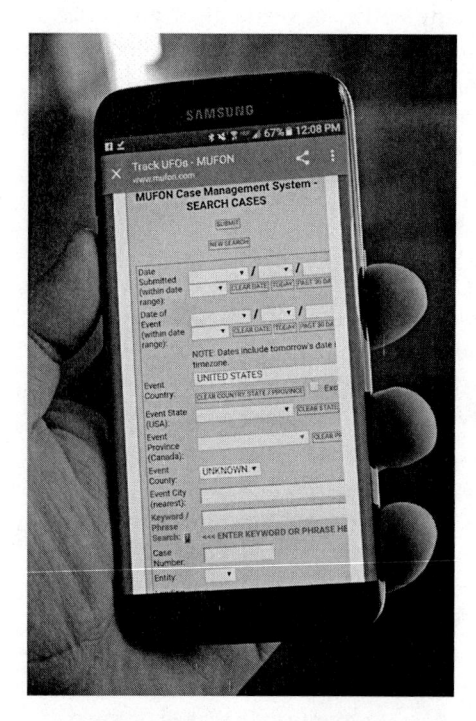

The MUFON database website allows anyone to search cases submitted to the Mutual UFO Network, including photographs and videos from witnesses. Photograph by Kenny Biddle.

"Scientificity"

ARIG websites that specifically noted or suggested a sense of *scientificity* did so in a number of ways. One way is reference to a systematic method or "the scientific method." Another is emphasis on gathering of objective measurements with specialized equipment. "Scientificity" describes the degree to which ARIGs addressed science—if they actually used the words "science" or "scientific" to describe their group or procedures or if they suggested they used science-related processes. Examining scientificity of ARIGs allowed comparison between ARIGs' understanding and use of science and scientific research with academic and professional scientific research.

Using Internet browser tools, I scanned each ARIG home and "about" pages for use of the words "science" and "scientific" with reference to their method, goals, mission, or process. If "science" or "scientific" was mentioned, I determined if the context was invoking or revoking science. If they were outwardly hostile to science, I counted scientificity as a "no." If the terms were used to positively describe a characteristic of the ARIG, scientificity was counted as "yes." Greater than half of the groups, 526 out of 1,000, were positive for scientificity. An additional two groups were more tenuous and used "quasi-" or "semi-" as a modifier in front of scientific. This cautiousness suggested that some ARIGS are aware of how different their activity is from academic science. Statements such as "this is not an exact science" soften the scientificity. There are those who clearly state their opinion that investigation into the paranormal, cryptozoology, or ufology is science. Many others will just reference the equipment they use as "scientific" and others strongly suggest scientificity by referencing scientific works or famous scientists, like Einstein or Tesla. A completely non-scientific approach was evident only on 19 sites, all of which carried religious connotations or advocated a strong metaphysical foundation and spiritual approach. Meanwhile, 408 of the sites did not specify or hint at if and how they used scientific methods, or if they thought their efforts were scientific.

Those ARIGs that strongly portray scientificity in their presentation to the public consider their subject to be an uncharted, ignored form of science and, therefore, conclude the scientific community is unjustly ignoring this field. Anyone who has examined the idea and believes there is "something to this" sees a preponderance of evidence for life after death, Bigfoot, possession, psychic powers, or unexplained things in the sky. In their view, the evidence is obvious and highly convincing. They tout their evidence as if adding photos, stories by eyewitnesses, and their own personal experiences will persuade the scientific community. In general, ARIGs do not exhibit knowledge of the scientific and scholarly history of exploration into fringe areas. Quotes from some scientific minded groups will illustrate their ambitious goals. Their missions included:

- "furthering the science" (In the Shadows Paranormal Project);
- "bringing science and paranormal together" (Peace of Mind Paranormal Society);
- "helping the scientific community embrace the world of unknown" (Eastern Kentucky Ghost Hunters—EKG Investigators); AND
- "compelling the scientific world to action" (Texla Cryptozoological Research Group).

2

The Paranormal in Popular Culture

Those who do not believe in paranormal claims are regularly astounded at how many people around them subscribe to paranormal thinking. Paranormal topics are presented in American popular culture as real, both in fictional works as well as "true accounts." In popular media and survey data, paranormal topics are not marginal; they are mainstream. Percentages from public polling show a substantial belief in paranormal phenomena. More than half the population at any time admits to accepting some mysterious concept as real. From 1990 to 2001, belief in haunted houses and ghosts each increased 13 percentage points—from 29 to 42%; communicating with someone who has died was up 10 percentage points to 28%; and extraterrestrial visitation was up 6 points to 33% according to Gallup polling.[1] Those numbers came down slightly in 2005 which was the last set of results available.[2] These fluctuations are not dramatic and point to a persistent belief. More than half the U.S. population subscribes to at least one paranormal or supernatural belief. It's possible (but unclear) if the high point in 2001 was due to the popularity of the paranormal in popular culture, though that hypothesis is reasonable.

Paranormal belief appears to be correlated with paranormal media. Support for paranormal media comes mainly from the audience that subscribes to that paranormal worldview. A 2009 Pew survey of over 2000 people showed 29% of people reported they have been in touch with the dead in some way and 18% experienced ghosts. In total, 65% of the population of adults express belief in or report having experience with at least one supernatural[3] phenomena (Pew Research Center 2009). Is this the result of psychic celebrity shows on TV and the several programs that relate personal stories of ghost encounters? A study of belief in ESP suggests that people are generally more willing to accept the truth of claims if others around them appear to accept them also (Ridolfo et al. 2010).

Paranormal topics occupy an uneasy spot as alternatives to mainstream religion and entertainment. In *Paranormal Media* (2010), Annette Hill, a media professor, charts the media representation of paranormal beliefs. Hill (no relation to author) says belief in spirits was immediately intertwined with the development of mass entertainment. Capitalization of those beliefs was maintained through pop culture, though they have undergone evolution beginning with the popularity of Spiritualism, a religion of Victorian times that arose in conjunction with the growth of scientific thinking and new technology. Today's paranormal belief is not so much about religion as about a desire for unique and extraordinary cultural, social, and emotional experiences. Spirits reflect socio-economic and political changes, changing morals and cultural imbalance (p. 35). UFOs grew from the military industrial complex and during times of fear of foreign invasion. Bigfoot represented natural existence. Hill argues that this evolution is coincident with changing worldviews and how we understand reality. Culture and paranormal beliefs are intimately connected and move together. A particularly pertinent point was that people enjoy seeing their beliefs played out in public situations.

In 2010, the book *Paranormal America* (Bader et al. 2010) was the first comprehensive look at people who are heavily invested in a paranormal belief system. *Paranormal America* relied on results from the Baylor Religion Survey to examine the less overtly religious landscape of U.S. paranormal beliefs relating to ghosts, astrology, UFOs, psychics, and Bigfoot. Bader et al. were surprised that the participants were not "marginal" members of society, describing them as "very normal people talking about a very strange subject." They also noted the oddness that the public is both interested in and repulsed by paranormal ideas—we are curious and feel compelled to subscribe to these emotional concepts, but at the same time may be concerned these beliefs render us irrational, gullible, or superstitious.

Several of the findings in *Paranormal America* echoed my own observations on the influence from television and the Internet in reinforcing and expanding beliefs into the American mainstream, making fringe ideas more palatable to the general population. Bader et al. also noted how some individuals and even municipalities capitalize on paranormal curiosity by promoting their mysteries as tourism or merchandising opportunities.

With two-thirds of Americans saying that they believe in at least one paranormal subject, we are faced with accepting that the paranormal is *normal* in the U.S. We can hardly label this subculture as "counter-culture" anymore (Dobry 2013).

Paranormal communities lack structure and stability, particularly in comparison to organized religion. Throughout *Paranormal America*, the researchers contrasted loose paranormal communities with religious communities, hypothesizing that perhaps people leave religious communities for various reasons and find a replacement in the paranormal scene. The relationship is not straightforward but they did differentiate two distinct styles of paranormal practice and belief. *Enlightenment* seekers are inwardly focused, interested in personal understanding and enrichment. *Discovery* paranormalists seek evidence of the phenomena that can be presented to the world. Discovery paranormalists are the core of ARIGs. We can further subdivide paranormalists based on other habits. Paranormal particularists may accept one phenomenon as real but reject another (Bader et al. 2010). They are more likely to focus seriously on one topic and not take another similarly categorized belief seriously at all. This is exemplified by Bigfoot "naturalists" who think of the entity as a physical flesh and blood creature. They typically want nothing to do with ghost or UFO believers. However, there are paranormal generalists that accept a wide range of paranormal ideas and tend to see interconnectedness among them. Some Bigfoot paranormalists accept exotic, fantastic ideas relating to UFOs, alternate dimensions, and spiritual or demonic association. Nonbelievers of all paranormal topics are the odd ones out when they suggest these reports are not mysterious but all have prosaic explanations.

Paranormal versus Supernatural

Paranormal themes are part of today's cultural landscape. What is popularly referred to as *paranormal* changes over time based on cultural and media representations (A. Hill 2010) but the word itself didn't always exist and, in the recent past, it meant something different. When I talk about "paranormal" I mean it in the context of today's subjects, what the participants believe to be paranormal. What exactly is that? *Paranormal* means "beside, above or beyond normal" (Baker & Nickell 1992). It is that which does not appear to fit into known mainstream science. A working definition for my purposes would be: those extraordinary phenomena perceived to defy explanation or not yet explained using current scientific understanding. Therefore, *paranormal* is exclusionary—all that which is not normal (Collins & Pinch 1982). Of course, that does rely on the degree of unusualness the experiencer perceived and ascribed to the experience. Paranormal is an alternative narrative to normal. "Perceived" is a critical distinction. That important concept—that a sighting or phenomenon is one or more persons' perception, not facts—is glossed over by casual investigators. Paranormal activity is presumed to exist by many people. But, that is not proof that it does exist.

Though some argue otherwise, I contend that *paranormal* should be distinguished from *supernatural*, and the distinction is considerable. *Paranormal* allows that we may yet discover a normal cause and redefine natural laws to accommodate the phenomena, or the phenomenon will one day come into the realm of established science. *Supernatural* does not suggest this. The supernatural cannot be examined under scientific processes since, by definition, natural rules do not apply which means all tests are moot. If we can't count on physical laws for supernatural causes, we can set no boundaries. As such, *paranormal* and *supernatural* are not synonyms. Many ARIGs, and the media, use these terms loosely. Paranormal events can appear to be *supernatural*, and vice versa, but the ultimate distinction is based on the root cause proposed—part of the natural world we can measure and test, or outside of natural laws, unpredictable, and, hence, outside science. *Supernatural* also presupposes that operation is dictated by some agent or force—a god, demon, angel. Belief in agents causing events is ingrained in the human brain; we naturally assume there is an underlying cause instead of accepting the vagaries of the universe (Goode 2000). So far, we don't have any solid cases that are conclusively supernatural. Believers in miracles (as a matter of faith rather than evidence) might dispute this but there has always been a potential alternate natural explanation for miracles. We'll leave open the possibility that a day may come when an entity that can suspend natural laws shows his actions to humans by defying the laws of physics repeatedly in front of us with no alternative explanations. Until then, logic and odds dictate we'd best rely on natural laws to continue, as they have, unabated. They've not failed us so far. While some critics insult those who believe in paranormal phenomenon and supernatural entities, the core causes for such beliefs are a normal part of the human condition. Yet, in society as well as in science, both paranormal and supernatural discourse occurs on the margins, outside the mainstream.

The term *paranormal* has evolved through time, expanding in scope and becoming muddled in meaning over the past few decades. Some *supernatural* topics became *paranormal* in the 1930s (Clarke 2012). In the 1960s through the 1980s, "the paranormal" was largely synonymous with psychic powers, ESP (extrasensory perception), telekinesis, and clairvoyance. Since the 1990s, it has expanded to include several subject areas, but its primary use has shifted to referencing ghosts and hauntings—as *paranormal activity*. This new usage may have to do with pop culture marketing and the decline of formal parapsychology research programs. The term has been co-opted to gather similarly peculiar and fringe topics under one label for ease of media consumption, whether that be for books, movies, or television shows. The *paranormal* is such a diverse term that it encompasses what people perceive both as real and as fiction and it can engender extremely strong emotion. The extent to which some pursue paranormal subjects are similar to that of political or religious fervor (Grinspoon & Persky 1972).

The generalizing of the use of *paranormal*, ironically, normalized reporting events as paranormal activity. Though the experiences are almost entirely associated in the public view with lower levels of education and income, the themes are ubiquitous. One is hard pressed to find a Ghost/ARIG that does not assume paranormal activity is real, and they claim to find it all the time! If they didn't come across what they consider to be paranormal activity on a regular basis, there is a good chance they would get frustrated and give up. Instead, they are rewarded via their own constructs in a self-fulfilling event. This process will be discussed in detail in Chapter 12.

Because of the negative connotation of the term *paranormal* associated with weird, fringe, or supernatural and therefore, unscientific, some ARIGs object to being associated with the term. Bigfoot, UFOs and other anomalous phenomena fit under the paranormal umbrella in that they do not currently align with any accepted scientific explanation and continue to be seen as unexplained and mysterious. The concept of ghosts is certainly paranormal. Cryptids that are claimed to be hiding from man and not subject to established concepts of wildlife biology, zoology, or evolution are outside normal (Loxton & Prothero 2013). UFOs that can't be identified or that reportedly move unlike any earthly craft, deviating from well-established parameters of nature, are also included under the "paranormal." As described earlier, I use *ARIGs* to accommodate the diversity in methods or how they choose to portray themselves (as naturalists, skeptics, or interested researchers). I chose not to officially label the groups as "paranormal investigators," but reserve "paranormal investigator" for those that deliberately seek out paranormal evidence. There are certainly many under the ARIG rubric who are paranormal investigators; not all operate this way and, as I later conclude, they don't have to.

"Paranormal" is a form of narrative structure—a way of packaging the world to emphasize the mysterious and secret. We see social bonding under this narrative (though opinions on it can be widely diverse) (Jenzen & Munt 2013). As popular culture content providers capitalized on public interest in the paranormal, primarily with reality television and associated events, paranormal investigation groups after 2000 promoted an image of the dedicated, sober, technological, scientifically-minded, and skeptical researcher. That's what they hoped the audience perceives.

Paranormal Themes in Pop Culture

Themes of UFOs, hauntings, and monsters are ubiquitous and universal in pop culture. Paranormal phenomena themselves have national and cultural origins and flavors. The history of ARIG groups of each category shows that the origin of some paranormal subjects began in other countries and were imported to the U.S. UFOs seem to be a particularly American invention that we exported to the world, yet distinct types of aliens emerged around the world. The ubiquitous "gray" alien with big black eyes evolved out of the American media in the 1960s before spreading globally. Paranormal media, especially television "discovery" shows, caught on strongly in the U.S. Some of these cultural ideas may not have originated in the U.S. but they were given a great boost or reimagined thanks to the American popularity of ghosts, huge ape-like wild men, red-eyed mystery animals, and gray humanoid alien abductors.

Most every American knows the iconic Frame 352 of the Patterson-Gimlin film purportedly showing an alleged striding Bigfoot. Many people know that there are local paranormal investigators that offer to come to your house just like on the TV shows. Too many observers call any light in the sky a UFO. Because society is acutely familiar with these concepts, there is a psychological effect at play that causes one report to lead to even more reports of a similar event. This "contagion of experiences" (Westrum 1977) reveals that the popularity of reporting of such events is strongly conditioned by social forces (Sagan & Page 1972). In the history of UFO sightings in particular, we see patterns of reports that are called "flaps." These "flaps" which also occur for sightings of Bigfoot or other unusual creatures (such as Mothman in 1966–67 West Virginia and Jersey Devil of 1909 in New Jersey and Pennsylvania), can have more to do with what people anticipate they will see than what they actually do see. During the 1990s when the *X-Files* television show was popular, UFOs, the extraterrestrial hypothesis, and ideas about government conspiracies were flourishing. Instead of a rash of reports suggesting that experience is genuinely happening, it could be that witnesses are interpreting observations in this anomalous context because that's the frame of mind they are in. There are many examples of contagion including episodes of mass hysteria where many people experience illness or panic. One famous example of contagion is that of an escaped red panda. In 1978, a clever red panda absconded from the Rotterdam (The Netherlands) Zoo. The zoo received calls reporting panda sightings for a year after the animal was found dead near the zoo very soon after the escape. The "red panda effect" means people see what they are told to see, not necessarily what is really there.[4]

A more recent example is that of creepy clown sightings that occurred in 2016.[5]

The same contagion of experience can be seen in ghost experiences where every noise or anomaly noted in a so-called "haunted" location will be automatically attributed to the legendary spirit(s). If we are primed with a story regarding what might be seen or felt, people are more likely to interpret any stimulus in accordance with that preconceived notion. Also, human nature compels us to play along for the sake of fitting in and having a group experience. Childs and Murray (2010) explain in their assessment of verbal descriptions by paranormal investigators that context is everything. When two or more people describe what happened, the narrative evolves to include and reinforce all persons' views but rolls out in a way suitable to the social conditions, not as a strictly objective account of what happened. The authors note: "Rather than remember events in ways that reflect best attempts at accurate recall, speakers provided accounts in ways that attended to particular interactional business." In other words, the speakers attend to the *function* of telling about the events. If the implicit goal is to present evidence suggestive of paranormal activity, the speakers will follow a pattern of telling that includes means to head off arguments or conflicts and using devices (unintentionally) that enhance their credibility and trustworthiness. Examples include avowal of prior skepticism (Lamont 2007), which is an assurance that the witness is not gullible, appeals to science and authority, and hesitancy to definitively proclaim the paranormal but letting the listener decide.

Critical counters to pro-paranormal themes are the skeptical (doubtful) or non-believer's views. Proper skepticism in terms of philosophy and science is described previously. Skepticism is also a culturally acceptable point of view. Hill emphasizes that skepticism is a "significant aspect of representation of the paranormal in pop culture" (A. Hill 2010: 59). The skeptic is thought of as the being the "rational thinker," a useful foil in a narrative where the experiences are described as extraordinary. The skeptic enhances the paranormalists' view by defining the end points on the belief spectrum—think Mulder versus Scully (*X-Files*) as the classic and greatly overused contrast, but some of us still think of Velma opposed to Shaggy from *Scooby Doo* as the ultimate representation of paranormal poles of belief. Paranormal TV shows today often include the "token skeptic" role (usually a scientist such as on *Finding Bigfoot)* and ARIGs often single out the "skeptical" member of their teams. Avowal of prior skepticism described above is such common behavior for ARIGs that it's nearly a defining characteristic. They will say they once were skeptical but now are convinced, providing a casual cross-check that suggests they have their wits about them and are not easily convinced.

Paranormal Media

Human societies routinely share stories of mystery, spirits, and monsters. Famous American writers Washington Irving and Edgar Allan Poe embraced supernatural themes early in U.S. history (Booker 2009) and wove this motif into the fabric of the new country's history. We are exposed to supernatural horror stories, myths, legends, and spiritual ideas from the time we are children and through adulthood. Mystery and sensationalism sells papers, movie tickets, or, in today's age, gets clicks. Society's fears, beliefs, and desires are turned into a revenue stream and cultural industries (A. Hill 2010; Clarke 2012).

The modern Spiritualist movement featuring psychic mediums and séances originated in the United States in the mid–1800s (Irwin 1989). Radio dramas and late night call-in shows helped propel popularity of the genre of ghost stories and science fiction. The Yeti and the Loch Ness Monster entered the scene in the 1920s and 30s. A post-war surge in unorthodox ideas introduced to the public to the likes of Immanuel Velikovsky, Ivan Sanderson, and Vincent Gaddis who popularized fringe ideas about the world. The term "UFO" was coined in 1950 by U.S. Air Force officer Edward Ruppelt as flying saucers were seen across the skies. Various other unorthodox scientific and occult ideas were discussed in the 1950s. Magazines for men and kids featured weird tales. Media coverage, especially television, was a key to spreading these new ideas and helped frame the public response (Clarke 2013). Bigfoot walked into the public consciousness in the late 1950s with the Patterson-Gimlin film that Patterson self-promoted across the country. By the end of the 1960s, there was an explosion of interest in strange phenomena (Thurs 2007) that led to paranormal media for mass consumption (Northcote 2007). Parapsychology, focused on untapped mind potential, established itself as a legitimate, but controversial, scientific field (Allison 1979; Collins & Pinch 1982; Hess 1993) from the 1930s through the 1970s. Astrology, psychics and "far out" New Age ideas fanned out into the culture.

Mass media has regularly featured strange content for its audience beginning with early newspapers reporting sea serpent sightings, poltergeist troubles, and strange things in the sky. But, the 1960s and 70s marked the proliferation of popular books and television programs on unexplained mysteries such as Bigfoot and lake monsters, the Bermuda Triangle, lost civilizations, haunted houses, and psychic abilities. With these fantastic claims came those determined to investigate the claims and nab the title of "expert." The *National Enquirer*, a tabloid that had existed for decades, put paranormal-themed exposés on their cover beginning around 1968 (Thurs 2007) when the counterculture movement was in full swing. A portion of the public became discontented and hostile towards the negative products of scientific progress like bombs, chemicals, and industrial waste. Alternatives to orthodox science, like ufology and parapsychology, grew popular at this time. In a snapshot from 1973, a survey showed that a whopping 95% of the American public was aware of what "UFO" meant (Denzler 2003) and many thought they were a threat.

Reports of "high strangeness" (as coined by J. Allen Hynek) were given the "X" label, which stuck. It means paranormal, supernatural, or "extremely" out of the ordinary (Thurs 2007). It's also the unknown variable in mathematics. The "X" factor made its most indelible mark with *The X Files* TV show, which ran from 1993 to 2002 and was resurrected again in 2016. This series was a key pop cultural product related to investigating the strange, and introduced many viewers to conspiracy ideas that were already established in remote corners of society.

The linkage between different fields of strange or weird studies into "paranormal" was forged by television shows like *In Search Of...* (1976–1982), *Unsolved Mysteries* (1987–2002, 2008) and *Sightings* (1992–1997), magazines such as *Fate*, and certain authors, most notably John Keel. Charles Fort (1874–1932) provided very early grist for the mill with his collected accounts of anomalies he called "damned data" because they were incongruent with scientific observations of the time. Fort's collections of anecdotes and accounts remain somewhat popular today, most notably through the *Fortean Times* magazine, billed at the "World's Best Source of Weird News." Fort would likely have been flabbergasted at the modern commercialism of weird news. Reporting of anomalies is big business as several news outlets now seek out and publicize strange stories and many Internet sites obtain huge numbers of visits by specializing in reporting anomalous news events (often without any investigation of the claim). Big-impact outlets such as Associated Press and *Huffington Post* have spun off specific branches of their social media stream or web space to feature oddities and anomalies.[6]

Indeed, paranormal topics have become so normal a part of popular culture that they are mundane. A vast array of products capitalize on ghosts, UFOs, and cryptids as selling points.[7] Casual interest in these subjects is acceptable, but it's still considered a bit odd to go Bigfoot hunting or try to spot UFOs in your spare time. Television and the Internet have lessened the stigma surrounding discussion of these topics in public arenas.

Paranormal TV

Television is a major influence on our culture and evolved rapidly into a primary means by which we learn about the world. The explosive growth

An example of a "mysterious shadow" created with a slow shutter speed by a subject pausing for a few seconds, and then walking out of frame. By stopping midstride, he reflected enough light to make an impression on the image. Photograph by Kenny Biddle.

since the 1980s of 24-hour cable channels, satellite service and the Internet has dramatically increased the need for content. Face it: a scientific process simply doesn't make for exciting television viewing. So reality is edited and dramatized to show forensic or scientific investigation in a superficial way to solve crimes or mysteries. Ambiguity, nuance, and complexity of the human experience are more difficult to depict than emotions like fear and hope. Educational programming was too dull so it was retooled into entertainment.

Prior to the reality-TV era, which began around 2000, shows like *In Search Of...* and *Unsolved Mysteries* portrayed concepts of the paranormal with an unnerving uncritical angle. *In Search Of...* had a documentary style but *Unsolved Mysteries* went further to provide dramatic recreations of purported events. Even though it commonly was left up to the viewers to "decide" what to make of these supposedly true stories, there is no doubt that these shows framed the stories with an effort to persuade the viewer to consider a paranormal interpretation. With the rise of reality television, "experience" rather than arguments or reasoned conclusions became the norm. Today's modern paranormal-themed shows are blatant about belief and enhance the persuasion with drama, night vision camera work, scary music, voice-overs and a narrative that conveniently ignores other interpretations (Harvey 2013). The "angle" is picked first and then all the supporting content is added. This is backwards from a scientific process where the investigator surveys the data first before coming up with a theory to explain it. Unfortunately, that's not a great format for a television show. Scientific ideas are not judged on their entertainment value (Mooney & Kirshenbaum 2009). Despite the illusion of scientific investigation, paranormal reality shows are over-simplified, dramatized, and sensationalized to appeal to an audience who desires to be thrilled and amused rather than well-informed. Inaccuracies creep in or are ignored. Reenactments of events, for example, are necessarily interpretation. Dramatic flourishes may be inserted that distort or misrepresent historical accuracy. Scientific processes and the capabilities of a scientist are distorted and incomplete, even dumbed-down to the point of absurdity. The role of investigator is edited and dramatized to the point of being cartoonish. Failures, mistakes, possible alternatives, and ambiguities are excluded from the story because of time limitations or perceived lack of interest. Viewers, who don't get the whole story, are encouraged to decide for themselves. This method is manipulative, requiring the viewer to accept that the TV presentation provided all they needed to know to make an informed decision. The results are of poor intellectual quality (Hufford 2001). Unsupported ideas are afforded a degree of respectability based on their entertainment value.

The paranormal was well-publicized but not legitimized. Skeptics and even members of the paranormal community argued that the paranormal TV boom promoted misinformation and confusion about paranormal topics. Regardless, television heavily influenced many future ARIGs to get involved in "the field," starting them off with a wrong impression of what is fact and what is unsupported speculation. Popular media is an important element in learning about culture (Mooney & Kirshenbaum 2009). Those eager to discover the paranormal world for themselves see everyday folks "investigating" on TV and think, "Hey, I could do that." Even the fictional *The X-Files* inspired would-be investigators to imitate FBI agents Fox Mulder and Dana Scully.

Paranormal non-fiction television in the 2000s was a hybrid of reality

TV and history documentaries (A. Hill 2010). Shows like *Most Haunted* in the UK, the first in a long line of similar shows, reflected and reinforced cultural trends of paranormal belief (Koven 2007). One result of their popularity and influence is the common assumption that any historic place must be haunted. Talk of resident ghosts or monsters became more prevalent among those who embraced the role of "hauntrepreneur"[8] to promote tourist income. The owners of several hotels, houses, historic buildings (especially prisons or hospitals), restaurants, and pubs capitalized on the exposure gained from paranormal television to draw more paying visitors eager for their own spooky experience.

The lucrative new wave of paranormal TV was exploited by cable networks specializing in human interest and curiosity about the world: A&E, Discovery, Travel Channel, History Channel and The Learning Channel. By airing supposedly fact-based programming related to hauntings, UFOs, and cryptozoology, these media outlets mainstreamed paranormal topics for people who had never been exposed to the ideas in a serious (seemingly non-fictional) way. Television propelled the activity of amateur research and investigation to an acceptable, even trendy, pastime through real depictions of investigators. The ARIGs will readily admit to television as being a major influence and an impetus to explore the paranormal on their terms. Brown's (2008) book *Ghost Hunters of New England* contained interviews with ghost hunters who cited the strong influence of paranormal-themed television shows such as *Sightings*, *Unsolved Mysteries* and, most often, *Ghost Hunters* on the SyFy network, featuring the crew of The Atlantic Paranormal Society (TAPS). ARIG members told Brown that TAPS opened the field of ghost investigation to the public and inspired many to form their own groups (pp. 85, 146). *Ghost Hunters* became a brand. Other popular investigators or writers built their own brands to capitalize on this new social niche of the average person as paranormal investigator. Paranormal belief was turned into a revenue stream for television networks and advertisers and was used by related parties to capitalize on the interest in weird things.

Proliferating in the 2000s, License (2016) refers to paranormal reality TV as "immersive theatre" as the participants were at an interesting location with a curious story and several props. The backstory (which was often fictionalized and never fact-checked by the investigators) was reinforced by attributing any anomalies found to the possibly fictional narrative.[9] The shows not only displayed a veneer of science, according to License, but a veneer of history. The backstories dictated what would be found and how it would be interpreted—not conducive to an objective investigation at all.

In 2013 alone, 19 paranormal series premiered on U.S. television, 23 in 2014, and 17 in 2015.[10] These high numbers mark a shift in the 2000s to high-volume, low-cost paranormal TV. Producers correctly banked on the idea that superficial and positive narratives of the paranormal would be more popular than complex and skeptical views. A good skeptical approach has rarely been attempted so we cannot compare if such a view would or wouldn't resonate with audiences. A typical pattern for production was this: once a key show was successful, spinoffs (or copycats) followed in order to carry the core audience. *Ghost Hunters* was the most popular of this genre. Over two million viewers (Seidman 2009) tuned in per episode to make this SyFy channel's most successful series. It spun off two other *Ghost Hunter* shows—*Ghost Hunters International* and *Ghost Hunters Academy*. The majority of paranormal TV presents a core formula: heroic investigators tackle a mysterious situation. The popularity of paranormal shows suggested that a subset of America was fascinated by the topic. Fans of these shows wanted to believe in something more than what the real world offered and, as with television in general, these shows provided an escape from one's own life. The do-it-yourself, everyman attitude encouraged audience participation, imitative actions, reactions, and behaviors. Social media and Internet forums were used effectively to build an audience and for discussion and increased promotion. As with sports teams, musical artists, or other television shows, being a staunch follower and fan of this or that paranormal show became a part of a person's identity.

Celebrating its 10th season on television in 2014, *Ghost Hunters* received harsh critical review that contrasted with the great ratings. Bruno Maddox (2009) observed that *Ghost Hunters* represented "how easily and thoroughly any humdrum existence can be transformed" by focusing on the mysterious. He also called the show "deeply stupid" with respect to the airs of "science" that they put on. But his observation hit on a critical component of ARIGs: do-it-yourself mystery hunting can transcend the drudgery of everyday life. Thanks to reality paranormal TV, mystery hunting became a culturally acceptable novelty not seen since the days of Nancy Drew and Scooby Doo (Sagan & Page 1972). Even though the shows were not designed to educate, people followed their lead as if they were.

Ask those who sell ghost hunting equipment and they will tell you the most popular device at the moment is the one seen on the ghost show that past week.[11] TV investigators are small-scale celebrities. Collectively, they are casually referred to and treated as "para-celebs" at conferences and public appearances. Once they reach the level of having their own show, their status is raised, and they acquire a fan base, but respect among their colleagues often dwindles. There is a clear sense that "selling out" for a TV show means leaving research integrity behind. Jealousy may be at play as well, as the popular para-celebrities will be paid to appear at events.[12]

Though some initially friendly networks for these mystery-mongering shows eventually became less welcoming and dropped series, others such as Destination America channel picked up the slack. Destination America, in 2016, seemed to be on the cusp of being all-paranormal all the time. It's curious to note the association with America to all this weirdness and mystery. Any place you go in the U.S. appears to have some spookiness attached.

The widespread effects of these reality-type paranormal shows on paranormal belief has not been well studied. It is still unclear exactly what buttons they push in people to compel them into investigating the phenomena themselves as opposed to enjoying investigation stories as entertainment. However, television has a tremendous impact on how we perceive the world and gage reality. It can also distort reality (Toumey 1996). Sconce (2000) suggests that if we see so much relatable "life" on TV we lose verifiable reference to reality. The barrier between television portrayal of real life and actual real life dissolves. Is this a cause of ARIGs imitating media? Paranormal investigator shows such as *Ghost Hunters* and *Finding Bigfoot* are examples of *hyperrealism* in television. The shows are an exaggeration of reality that do not accurately depict what these people would do without the camera on them. It is a manufactured reality, an enhanced, more dramatic, condensed, and concentrated version of reality. The hazard that viewers may mistake hyperreality for reality is not just a possibility … it happens! People visit paranormal hotspots looking for their own experiences ("ostension")—sometimes resulting in a fuming property owner. Amateurs on these shows have become role models that viewers emulate in their own real-life situations. The television depiction of "ghosts," "hauntings," or "Bigfoot encounters" have usurped the narrative so that what is seen on TV is now assumed to be in the environment outside of television. It's real, they say, "I saw it on TV." Thus, ghosts, Bigfoot, and alien abductions are defined by a fictionalized or invented TV narrative. It's a dangerous precedent to confuse facts with fiction.

Science-themed television shows may boost the popularity of science. Or, they may boost the popularity of "sciencey" ideas as there is a disconnect between what is shown to the viewer and the actual substance and relationships present in science (Denzler 2003). Mooney and Kirschenbaum (2009) argue that television portrayal of "science" can potentially be dangerous, such as in cases of medical information provided and decisions made from nonexperts. Because of the television business model of homogenization of content, we are flooded with many bad portrayals of fake reality. If we get our ideas about reality from the media, will people exposed to bad science, or dramatized, unrealistic science, eventually think science is just drama with gadgets? Does the trend of inaccurate scientific processes on television negatively affect the public's ability to make informed judgments? Will they support policy decisions based on pseudoscience instead of those based on good data and reasonable conclusions? Good decisions can't come from a culture deliberately misinformed about the world around them. But right now, that scenario is playing out every programming hour.

Media Contribute to Beliefs

The 1984 movie *Ghostbusters* gave a boost to the idea of real life ghosthunting even though it was pure fiction (Mayer 2013) Its influence on the protocols and process of paranormal investigation is debatable (considering it was a fictional comedy) but there are sadly some who really do think Egon Spengler's proton pack is a real device.[13] The movie had key elements such as use of technology, real references to parapsychology, demonic possession, and even good old ectoplasm—combining old ideas into a modern tale. Thirty years later, the movie is iconic. When we think of hunting ghosts, we can't help but recall the idea of the Ghostbusters revving up their proton packs with explicit instructions to not cross the streams. Parapsychologist Ciaran O'Keefe admitted that it was a huge influence on his career choice.[14] Ghostbusters made paranormal investigation (mainly by men until the women were given center stage in 2016) cool (Mayer 2013).

Leary and Butler (2015) cite that the 2005 movie *White Noise* exponentially increased the interest in electronic communication with the dead. Ghost hunting TV shows later capitalized on this. A 1972 BBC television movie, *The Stone Tape*, adapted and popularized ideas suggested by some paranormal writers regarding ideas that psychic energy released from past events can be captured in physical fabric of building materials and replayed like a magnetic tape—the dominant recording media of the 1970s.

A 2006 survey cited by Clarke (2013) noted that 75% of respondents obtained whatever information they had on anomalous lights in the sky (including potential "UFOs") from *fictional* TV shows and films. Thanks Steven Spielberg!

Modern paranormal content taps into these popular public sentiments perpetuating unsupported claims and misunderstanding about the role science plays in investigating them. The scientific concepts in *Ghostbusters* became normalized and the pretend science felt intuitive. Many beliefs are accepted because they feel intuitively right even if the supporting science is completely wrong (Blancke et al. 2016).

The popularity of pop culture paranormal is enhanced by the pleasure people feel in seeing their beliefs represented for everyone to see. Belief in something is connected to one's personality and needs in life. Belief is easy to reinforce, and hard to shake. It provides us with a sense of control and a framework for which to define our world (Belz & Fach 2015). When we see visual media that reinforce that belief, it becomes part of our worldview, even if it is fictional. Sometimes, humans respond so powerfully to fiction, it is as if it is real. Then, we may substitute that fiction and respond to it as to reality (Bartholomew 2012). Hill (A. Hill 2010) considers reality television shows about paranormal investigation to be "second-order experience." It's real enough to matter.

Factual or fictional, documentary or summer blockbuster, Americans absorb paranormal ideas and incorporate them into their worldview. The more paranormal themes are presented, the more commonplace and less strange they seem. Self-proclaimed experts become authorities on entities that no one has shown exist! Instead of passive consumers of the media, a more thoughtful approach would be to think and question what we are consuming, why we like it, and how it affects us. In some respects, it's not just a TV show or movie, it's an important part of our identity and worldview.

The Internet

For niche communities and alternative beliefs, the advent of the Internet changed everything. Increased access outside of academic and government use and the introduction of publicly available web browsers in the early to mid–90s allowed anyone to self-publish thoughts, opinions, photos, videos, discussions, and diatribes. Internet participants could exist under another identity, build their own sub-community, market themselves under their own brands, and gain attention and legitimacy with an ease like never before. It was, literally, a game-changer (Haythornthwaite & Kendell 2010).

As a great equalizer (Mims 1999) for widely variant points of view, perhaps the Internet has been *too* great. Internet search engines, the main gateway to information of personal interest, give equal weight to scientifically unequal ideas. Popularity is not dependent upon veracity. It seems self-evident that reasonable, moderate, and nuanced explanations are ignored as boring. They are no match for wild and fantastic stories and claims. UFO sightings and associated conspiracy ideas roared back into life again thanks to the Internet where alternative figures like Alex Jones, America's most popular conspiracy theorist, became media machines.[15] *Coast to*

Coast AM, a late-night radio show about fringe topics, became a popular interview forum for paranormal and conspiracy figures. This audience readily transitioned to the Internet where countless paranormal-themed discussion forums allow anyone to post their ideas and anyone else to consider or challenge it. To be a self-styled "expert" in the fringe topics is a strange parallel to being considered a scientific expert. Paranormal media figures compare or contrast themselves to scientists and, increasingly, media producers and providers seek out these self-styled experts instead of academic researchers.

Mass democratization of publishing removed gatekeepers and allowed anyone, not just professionals, to provide their contributions worldwide. Traditional forms of publication required editors and standards and, prior to the Internet, marginal ideas had a difficult time getting exposure (Shirky 2008). But now, it's easy to find and associate with those who share interests and to deliberately exclude opposing perspectives. Increased media attention to these subjects especially across the Internet has resulted in increased interest in the paranormal (Bader et al. 2010; Molle & Bader 2013; Gibson et al. 2009).

Also widely available now at little to no cost are *tons* of old records, databases, newspapers, and collected accounts. Gone are the days of getting motion sickness and headaches from scrolling through tiny print on a microfiche screen display. Social media tools have enabled a spread of information unprecedented in human history, and software and applications let us connect immediately with individuals and groups of people around the world in real time. Social networks were a boon to fringe subjects. MySpace, which established a network of users with real-time announcements and heavy use of images and sound, was the first place for ARIGs to set up their own webpage without the need for advanced computer knowledge. MySpace was already being supplanted by Facebook at the time I examined ARIG websites, but those that still existed typically contained grinning and flaming skulls, spooky music, crude animation graphics, and auto-play videos. The sites Meetup and Facebook broadcasted the time and location of local get-togethers that drew people to events.

People naturally wish to share their stories and seek out those with similar experiences who will reinforce their own beliefs and conclusions. The Internet has made this process a smooth pathway, coalescing believers in these *X-Files* subjects into online communities that function almost like a family. No idea is too "out there," and is nurtured through mutual support and encouragement. Just creation and sharing of an idea gives it life. It's made real by reinforcement and accepted because of the increased and repetitive exposure in our culture. The classic example would be the story of the Slenderman—a fictional evil creature from a creative forum that became so popular that people claim to see it and consider it real.[16]

Familiarity breeds acceptance. Continued exposure to these unique views makes them sound less weird the next time we hear them. Little steps from strange to stranger can take us very far away from where we started. The stigma of belief in the subjects of UFOs, Bigfoot, monsters, ghosts, and anomalies became normalized with assistance of the Internet. The paranormal feels familiar and seemingly plausible because we've heard so many stories that we assume there must be something to them.

The next three chapters provide a history of each ARIG focus area showing changing attitudes of the scientific community and the public through time.

3

Ghost Hunters and Paranormal Investigators

Paranormal research didn't begin with the *Ghost Hunters* TAPS crew in the 1990s. Ghost hunting has a long history as hobby, entertainment, and serious research topic. Ghosts are a way to explore the question of life after death. The early researchers into mediumship or spirit communication were considered "ghost hunters."

The first official media circus surrounding a haunted house was the Cock Lane ghost of 1762 London. Curious locals and visitors gathered at the apartment hoping to participate in a séance with "Scratching Fanny," the ghost that communicated through knocks that she was murdered. A committee was organized to investigate the ghost, eventually concluding it was a fraud (Clarke 2012).

Ghost hunting "flash mobs" were popular during this time too as several hundred people would gather around a building claimed to have spirit activity and watch for any hint of movement inside. These mobs of thrill-seeking citizens were often poor and working class looking for some excitement (Clarke 2012).

The oldest group with an interest in ghostly activity is The Ghost Club. It still exists today. Formed in 1862 in the U.K. to investigate claims of ghosts and hauntings, members included such socially esteemed people as academics from Cambridge University, clergy, and even author Charles Dickens. After a decline, it resurrected in 1882 as a secretive social club with occult leanings, with many notable members of the literary and scholarly community.

Psychical Research

Also in 1882 London, the Society of Psychical Research (SPR) convened for the first time with some overlap of members with The Ghost Club. SPR's goal was to bring scientific techniques and attitudes to psychical or spiritual claims. It was the first organization to scientifically and systematically study the reported phenomena of hauntings, poltergeists, and spirit activity. The SPR was made up of professional men and scientists, whereas The Ghost Club was more casual and advanced a belief in psychic power. The aim of the SPR was to find proof of life after death by studying reports of apparitions, haunted houses, and thought transference (telepathy), and to investigate self-proclaimed psychic mediums. Conventional science was not examining these remarkable phenomena at the time so the SPR took up that task. These scientists applied the scientific knowledge of that time regarding electricity, magnetism, and photography which was the first major application of apparatus to this subject area, and it would grow. The early SPR sentiment was that science would reveal much about the unseen world and possibly the keys to communicating with the dead. Belief in ghosts was close to being respectable as it appeared science was about to reveal their existence (Clarke 2012). A related organization was eventually established in the U.S.—the American Society of Psychical Research (ASPR), founded in 1885. The investigators were also trained scientists and their findings were published in their journals. Many prominent members of these organizations concluded that there was ample proof of life after death already (Lyons 2009). Opponents disagreed on the quality of evidence and questioned the susceptibility of these researchers to be fooled by fakers. There is a rich history of individuals, experiments and investigations that occurred during this vibrant time of psychic exploration. Legitimate scientific exchange took place in the pages of esteemed journals like *Science* and *Nature*. The hope was that scientific standards could elevate the field of study to accepted science. While many scientists still thought it was nonsense, many learned men (there were very few women involved except as mediums) were convinced that proof of communication with the dead was plausible and worth pursuing. However, many investigations uncovered fraudulent mediums or hoaxing. Disagreements erupted between SPR members about how strictly to apply scientific constraints in investigations or to allow looseness that may be more conducive to manifestation of the effect (and allow for ease of hoaxing) (Blum 2006).

Repeated exposure of frauds, also highlighted in the media by famous debunking magician Harry Houdini, further tarnished the already questionable reputation of the field of psychical research.

Not much has changed regarding psychics and paranormal phenomena as several proponents remain convinced and the scientific community in general is not. The SPR and ASPR still exist today, though the latter is basically defunct. The Journal of the SPR is still active, largely a legacy of the early intellectual branch of psychical research, almost entirely separate from TV-type ghost hunters (Jenzen & Munt 2013). There is a small cadre of academic researchers that still study extraordinary claims though it is not comparable to the process of ARIG's ghost investigations. (See Maher 2013.)

The majority of ghost/ARIGs are not familiar with the groundwork that has already been established by professional psychologists, historians, magicians, and physicists of The Ghost Club, the SPR, and ASPR. They obtain their foundation from modern media, not journals or historic sources.

Parapsychology

As the study of interactions between living things and the external environment that appear to operate outside of the known physical laws of nature, parapsychology remains as an academic science decidedly on the fringe. Once embraced by the American Association for the Advancement of Science (AAAS, publisher of *Science* journal), parapsychology has dwindled in size, influence, and credibility but coursework is still offered by dozens of genuine academic institutions. In the latest review of the state of the science, *Parapsychology: A Handbook for the 21st Century*, Cardeña et al. (2015) admit that parapsychology is considered a "borderline" area of

Ghost World paranormal conference in 2007, Gettysburg, Pennsylvania. Photograph by Kenny Biddle.

scientific research. Academic parapsychology encompasses areas of anomalous cognition including telepathy, clairvoyance and precognition, psychokinesis (mind interacting with matter at a distance), as well as the concept of survival of consciousness after physical death.[1] In contrast to the efforts of ghost/ARIGs to prove ghosts and hauntings are genuine, parapsychology as a discipline produces quantitative, experimental results published in scientific journals. Von Lucadou and Wald (2014) state that parapsychology (a term coined by Max Dessoir in 1889) has been devalued due to the "quacks and charlatans" enabled by the media who claim to be parapsychologists. Again, we see further support for the explanation that high level of belief in ghosts and hauntings resulted in a niche for self-styled consultants of the paranormal. The non-academic ARIGs use simple methods that provide immediate answers in contrast to the laboratory testing and statistical results of parapsychology.

From the start, parapsychology emphasized laboratory-centered experiments. Investigation of "spontaneous cases," those that happen in everyday life such as reports of hauntings and poltergeists, were not pursued as much. Louisa Rhine, the parapsychologist wife of Dr. J.B. Rhine who is considered the pioneer in the field, thought that spontaneous cases were too difficult due to errors in witness testimony. Weak subjective evidence was inferior to hard data obtained in the lab. The Rhines developed controlled experiments attempting to address the questions that arose from spontaneous cases (Williams et al. 2015). Spontaneous cases, if documented well, can reveal a pattern and suggest lines of further inquiry but no useful standards to investigate them has filtered down to all types of ARIGs who do not document their cases well enough for others to use in any meaningful way.

Researchers today still remain curious about the hypothesis of survival of consciousness after death and several institutions continue experiments. The Rhine research center (Institute of Parapsychology in North Carolina) is still active. Three other parapsychology research institutions that conduct experimentation include the privately-funded Windbridge Institute, the University of Virginia Division of Perceptual Studies, and the Koestler Parapsychology Unit at Edinburgh University in Scotland. Forty-three universities offer some form of formal education in parapsychology or anomalistic psychology. Thirty offer degree programs. The Parapsychological Association[2] and the Parapsychology Foundation, Inc.[3] are professional societies that provide grants and support scientific exploration of psychical research. These institutions have high research standards and maintain published works and databases of experimental results. I found no obvious cooperation between these institutes and ARIGs, though individual scholars may be involved with some ARIG cases.

Outspoken skeptics of parapsychology have had a profound influence on the field which took criticism seriously, enacting changes in procedures, and collaborating with mainstream psychologists for experiments and analysis. Today, results of experiments to test anomalous cognition are statistics-intensive. The general conclusion is that the psychic effect (or *psi*) is small, but real. However, the effects have not been demonstrated to be related to any paranormal cause. There remains no comprehensive agreement or model for what psi is and how it works, no ideas acceptable to explain hauntings, and no irrefutable evidence of ghosts (McCue 2002; Maher 2015). Speculative ideas about quantum effects, electromagnetic fields, tectonic strain, and stored psychic "energy" in materials with roots in academic parapsychology (or its precursor, psychical research) have been picked up by amateur paranormal researchers. Revealed in Cardeña et al. (2015) is the contrast between academic parapsychology and ARIG activity. The amateurs embrace EMF meters, consider electronic voice phenomenon (EVP) to be a primary form of evidence, and attempt to capture visuals of apparitions, whereas parapsychologists downplay these activities. They even chastise amateurs for their enthusiasm in evoking quantum mechanics as explanations for ghosts as this branch of physics is not amenable to amateur research (Miller 2015). The ghost/ARIGs' ideas about moon phases, solar influence, and space weather (geomagnetic storms) are not considered reputable by parapsychologists. Chapter 25 in Cardeña et al. (Maher 2015) provides the parapsychology view of ghost science, which is not subscription to the stone or water tape ideas of environmental recording so popular with modern ghost hunters today. Parapsychologists are uninterested in the amateurs' methods. The disconnect is stark between scientists studying seemingly paranormal activity and amateurs who say they experience it. Yet, there are those paranormal researchers who, regardless of any formal training, consider themselves lay parapsychologists (Childs & Murray 2010).

Harry Price

Most of those in the new wave of ghost/ARIG participants are not familiar with the substantial intellectual history of psychical research and the critical exposés by professional psychologists, historians, physicists, and magicians. The first media-savvy "ghost hunter" was Harry Price (1881–1948). Price was the "ghost showman" in the 1930s British press and the author of *Confessions of a Ghost Hunter*, a compilation of his adventures. A man of many interests, including science and technology, drama, and stage magic, eventually becoming a paranormal investigator, he joined the

SPR in 1920 and participated in restructuring The Ghost Club as more of a social group. Price tested mediums, exposed spirit photographers as frauds, and discovered the secret of making ghostly ectoplasm from egg whites. The debunking angle took a back seat as Price developed the first ghost hunter's kit consisting of flashlights, candles, notebook, and brandy (for nerves). He was not beloved by other experts but was popular with the media, placing himself as the investigator at the center of the story and making live radio broadcasts from "haunted houses" (Clarke 2012). Price, independently wealthy and free to indulge his interests, popularized ghost hunting as a form of entertainment and was influential in investigators' use of apparatus on an investigation (Clarke 2012). His most famous investigation, which he published, was of Borley Rectory, the "most haunted house in England." Price has been accused of omitting and suppressing information, of contradictions and inconsistencies, and possibly even hoaxing, in shameless promotion of himself (Davies 2007; Wiseman 2011) and Borley (Hines 2003). Strict British libel laws made such criticisms difficult during Price's life and ultimately limited exposure of skeptical suspicion. Alan Murdie, currently of The Ghost Club, admits Price had a hot temper and big ego and was a publicity hound.[4]

Scant few of today's amateur paranormalists know much about The Ghost Club, SPR, ASPR, or the popularity and techniques of Harry Price and other famous ghost hunters and busters of that era. They are oddly oblivious to the long history of serious scientific consideration of paranormal activity and the parade of personalities that took on the claims of evidence. There is a mother lode of written investigations and case studies from this time in the U.K. and U.S., stored in the archives of the SPR and ASPR, when psychical research was serious work for a few intrepid scientists.

The phase of modern ghost hunting began its rise in popularity in the 1970s. A very early organization was the Ghost Research Society (1977) that is still around today led by Loyd Auerbach. As of 2016, the Ghost Research Society website is typical of ARIG websites with auto-play music, a special ghost cursor to track your mouse movements, a black background festooned with advertisements.[5]

What Do Ghost/ARIGs Do?

Ghost "hunting" is a controversial term. It not only implies that there are ghosts to find, but it suggests a dispatching of the quarry at the end. Many ghost/ARIGs reject the term due to the inevitable association with the television ghost hunters and generally are more amenable to the term "paranormal investigators."

We have no working definition of ghosts. Ghosts themselves have changed appearances with the times (Finucane 1996). So "What is a ghost?" is a fundamental concept to consider. As ghosts are not recognized to exist other than as human constructs by most of mainstream science, seeking ghosts will differ from typical forms of research and investigation. Investigators who do not believe in the literal existence of ghosts will likely approach them from a literary, historical, cultural, or psychological perspective, as ideas or metaphors. For those who accept ghosts to be more tangible than a metaphor, they are assumed to be "remnant energy" of a dead person (or animal). Or are they souls? An electromagnetic recording of past events? A demon or the devil himself? Such ambiguity about what they are results in confused and unsettled plans for collecting data regarding ghosts. Ghost/ARIGs display this sense of impreciseness in their research, resorting to the use of dozens of widely varying techniques to assemble evidence.

Ghost/ARIGs most often investigate private residences by request, old buildings, historic structures, and places where deaths have occurred such as accident scenes, murder locations, hospitals, prisons, and military landmarks. Surprisingly, they still occasionally like to tiptoe around cemeteries—a location for investigation which remains a bit confusing to me. Why would ghosts haunt a location where no one died? Was it because graveyards are morbid places, oozing with the atmosphere of despair and sadness? Cemeteries have long been destinations for legend trippers and teens to display bravado in confronting scary claims. Atmosphere is crucial, as any good storyteller knows. Some people insist that cemeteries are portals to the afterlife and thus ghosts will be found wandering through as if it was a train station.[6] Famous ghost-storyteller Hanz Holzer popularized the idea that native burial grounds were a source of hauntings and believed that ghosts were a surviving emotional memory (Clarke 2012).

Historic sites, including cemeteries, in constant need of funds for maintenance, will rent out facilities for a night to paranormal investigation groups or allow special tours and events. Though considered more a "fun" event than a serious investigation, the sites (such as Fort Mifflin described in the Introduction) are recognized by many in the paranormal research community to be genuinely haunted. Investigators will try out new equipment, compare notes, and bring in guests or new people for the first time in these designated paranormal spaces.

Mayer (2013) cites three factors in the rise of ghost hunting organizations—television, the Internet, and the ease of obtaining equipment. He says that today's ghosts are akin to a problem with your plumbing: you notice it, call in the "experts" who learned on the job to investigate the issue, and then fix it with their tools. He notes that this idea of ghost "hunters" as a "service sector job" is being adopted by other countries, namely Germany, and it is popular in Italy (Molle & Bader 2013). In the survey of ghost-focused groups by Duffy (2012), information from up to 113 different groups was incorporated. Responses regarding training and education showed that no specific requirements were set for members to have before joining. If training was done at all, it was through the group itself. Having experience in science and research methods was cited for less than 5 percent of the groups. Also revealed in this survey were some demographics—most groups start out with about four members and build to a median of eight; the age range is generally between 20 and 40 years old. They conduct most of their investigations (about two per month) at private residences after they are contacted by the owners regarding possible paranormal issues. About half the time they say they find "paranormal activity" is occurring, but, contradictorily, consider it somewhat rare. About half the groups say that they have encountered a client who probably needs psychiatric help, and only a small percentage of groups will go as far as to consider an exorcism. Only about a quarter of groups say they registered as a nonprofit or a business. And, they consider their group function as mainly investigation and research (62%) but also to help people (58%) and to educate (31%). They will officially state that the main goals of their group are helping people and conducting research (94% of the time both goals were identified), but 72% say the goal is related to their own personal questions about the paranormal; 44% say the social aspect and fun of the activity is a main goal. Some 21% state their goal as getting a TV appearance, which is appreciable and bears remarking upon as distinct from the stated function of the activity. Eighteen percent have a goal of turning this group into a full business. While this is a small sample, these characterizations were consistent with my observations as well and, in my opinion, they present an accurate cross-section of ghost/ARIGs motivation.

As Seen on TV

The popularity of the 21st century ghost hunter can be directly linked to two television programs—*Most Haunted* in the U.K. (2002) and *Ghost Hunters* in the U.S. (2004). These shows were unique at the time and massively popular. There were antecedents: many ARIG participants have cited shows like *In Search Of...* (1976) and *Unsolved Mysteries* (1987) (both U.S.)

as inspiring or influential in their interest in the unknown. The 2000s saw an explosion of available channels and paranormal programming. *Ghost Hunters* was the catalyst for the exponential growth of both paranormal TV and of ARIGs in the U.S. (Brown 2008; Krulos 2015). They solidified the methodology widely adopted by ARIGs consisting of visiting people's houses, setting up equipment, and staying into the night recording anomalies. Even though paranormal investigation groups decry the TAPS methods and sell-out attitude, the number of these groups would not be in the thousands around the world if not for Grant and Jason, the Roto-Rooter plumbers-turned-celebrity-ghost hunters. They represent the "everyman hero." This same narrative resonates with those who become "monster hunters" and spot UFOs. Portrayed as "real life" situations on television, paranormal investigators dissolved the barrier between stories of the paranormal and actual experiences of the paranormal (Sconce 2000).

Women's interests in paranormal topics were not neglected by television producers. *Most Haunted* had a female host and was apparently developed specifically for a female audience. According to Hill (A. Hill 2010), a British network executive noticed that women's magazines had articles about the paranormal, specifically, psychic claims. For some of the audience, this show is sheer entertainment. For others, it rings true, emphasizing intuition and psychic sense. The viewer is left to decide, with very little help from the content, what to take away from the depiction. Several popular paranormal-themed shows today feature women, usually as the psychic-medium, a role often assumed by women since the days of Spiritualism.

The popular show *Paranormal State* (2006), aimed at young adults, was based on a group of collegians who formed their own investigation club at Pennsylvania State University. The travel-oriented *Ghost Adventures* (2008) featured a young, hip male host who made urban exploring sound like a fun time with your "bros." The parade of paranormal reality TV shows at times aimed directly at specific audiences including women, kids, celebrity-seekers, Latinos and African Americans.[7] The emphasis on appealing to specific audiences instead of on the nature or results of investigations, suggests how importance social interaction and identity formation is as part of ARIG participation. Clarke (2012) considers today's paranormal TV explosion as a modern day Victorian-era ghost hunting flash mob (p. 179). We all gather to see and collectively share in the real-life drama. Witness' accounts used to create and justify media stories and TV shows feed ARIGs' belief that "there must be something to it." But not many dig into the details to see if the core claims and evidence have merit, they just accept them, as does the public audience. Thanks to the proliferation of these stories, there appears to be an awful lot of "smoke" that leads people to conclude "fire." Instead, a cultural interpretation makes a lot more sense but is far more complicated and less TV-friendly; the truth is not a single entity or monster, but a complex interaction of perception, reaction, and cultural ideas that keeps changing and spreading through society (Loxton & Prothero 2013).

ARIGs unequivocally present a media-driven model of science. They admit they are inspired by these TV shows of "people making discoveries about spirits" (Ghost Hunters Guild) and that they "amassed great arsenal of equipment and run our team like the professionals you can see on the ghost hunter TV shows" (Wisconsin Area Ghost Investigation Society). They admit to the practice of taking notes from TV shows (Southern Spooks) and state they follow TAPS and *Paranormal State* TV shows in their procedures (Virginia Paranormal Organization of Research—VAPOR). Their idea of an investigation is "where you go to a location that is already haunted and set up equipment to search for results" (Mohawk Valley Ghost Hunters). Many use the term "reveal" as used on *Ghost Hunters*, to describe a quasi-formal discussion of evidence with the client. There is no doubt that ghost-hunting TV shows were the impetus for investigation into hauntings as serious leisure in the U.S.

Affiliations and Education

ARIG leaders will offer training classes for "ghost hunting" or paranormal investigation to the local community, particularly around Halloween. The classes range from free introductions to multi-day seminars and hands-on investigations that cost more than $100 per person. Paranormal presentations are given at local libraries, community centers or at a college or university. Small colleges will offer non-credit enrichment or continuing education courses for a small fee. Metaphysical institutions will also offer courses, even degree programs, online. These accomplishments have limited cachet and are not practically useful, though they are advertised as being a springboard to a career in the paranormal. Anyone with an interest can participate; there are no prerequisites or special skills needed. Occasionally a group will provide training and awards certifications for completion of their classes. The American Ghost Society offered a home study course as a prerequisite to joining the society. Some ARIGs find these courses of study to be an asset while others decry such programs, even going as far to label them scams.

Student clubs at major universities will sponsor talks by well-known paranormalists, or a ghost club will be formed on campus. Affiliation with an educational facility is highly desirable by ARIGs even if the institution is *not* endorsing the actions. A notable example of this was the Paranormal Research Society formed by Ryan Buell in 2001 when he was a student at Pennsylvania State University. The university did not endorse the connection between the club and his later TV show, *Paranormal State*. But the appearance of an association, as it remained an official student club at the university, was golden for Buell and the producers of the television show.

With few exceptions, ghost investigators are disconnected from the academic community. A university degree from the scant few academic parapsychology programs around the world is not viable for the majority of amateur paranormal researchers as this is their hobby, not a vocation or source of income. The lack of affiliation with professional scientists is a key difference between ARIGs and projects of public or citizen science since this work is not fed back to experts who check and publish these data and results.

There is a spectrum of opinions about the right and wrong way to be a paranormal investigator specializing in ghosts and hauntings. Popularity via the media introduced money and fame into what was previously a hobby. The result was strong resentment and animosity between groups. In the past few years, some groups promoted "paranormal unity," a ceasing of criticism and hostility between groups and unification under a common goal of research and investigation (Krulos 2015). Rather than minimizing strife, the effort stifled valid criticism of poor methods as ghost/ARIGs embraced an "anything goes" attitude—the opposite of a scientific, logical, or disciplined approach. Animosity between groups for notoriety and respect was greatly amplified by communication on the Internet, local competition for clients, and the frenzy to get the attention of a television production company. These conditions did nothing to encourage adoption of careful research and investigation standards.

4

Seeking Monsters: Bigfoot and Other Cryptids

Cryptozoologists believe there are hidden animals out there that can be found if we increase and improve our search of the wilderness. These mystery animals, called "cryptids," are the subject of legends around the world. They come to life in many eyewitness accounts, yet are most excellent at keeping themselves out of sight, except when they are dramatically seen—a strange contradiction.

Historical Zoology

Human interest in curious animals has always existed. We can fairly label Aristotle as the first zoologist, at least he tried to be systematic about animals beyond their practical use to people. Early naturalists got things wrong about the reports of animals they heard from around the world. In Willy Ley's wonderful *Dawn of Zoology* (1968), he traces the development of understanding about animals. In the Age of Discovery, many European explorers and missionaries repeated stories of strange creatures but did not investigate to judge their veracity (Regal 2011). Naturalists of the 16th and 17th centuries readily considered reports of all animals and "monstrous" humans. These reports were compiled into "bestiaries" from which the information was copied and accepted as representative. As travel became easier, explorers were able to bring back specimens of amazing creatures they had heard about from the natives. The specimens sparked debates about whether such strange creatures were real and where they fit into the order of life.

The New England shores were the location of a rash of sightings of sea monsters in mid–1800s (Loxton & Prothero 2013; Lyons 2009). Mysterious "sea serpents" were the world's first cryptids studied by scientists (though they weren't tagged with the name "cryptid" until 1983.[1] Rumors of giant squid, octopi and snake-like creatures were around since humans took to the sea. Mysterious underwater creatures were more plausible due to the vastness and poorly explored ocean depths. Sea serpent reports were so frequent and provided by respected officers and sailors that scientists engaged in professional debate about their existence (Loxton & Prothero 2013; Lyons 2009). The study of UFOs followed a similar trajectory into scientific debate after pilots reported unknown objects.

The break between acceptance and rejection of monsters began in the mid–19th century when fossil finds of prehistoric sea reptiles captured the public interest. Lyons (2009) details the extensive interest of the "gentlemen scientists" at the time who argued not only about the reality of sea serpents but about the process of how they should be examined. Some scientists, like Louis Agassiz, thought we would find a living representative of these creatures. Trailblazing geologist Charles Lyell privately collected sighting reports of sea serpents. If such beasts were found, he believed it would lend support to his theory of steady-state geology called "uniformitarianism." The predominant British scientist of the mid–19th century, Richard Owen, was skeptical of sea serpents because the evidence was solely from eyewitness reports, lacking any modern existing physical remains. Owen disregarded the testimonials collected by scientific committees and individuals, even experienced seamen. He argued that a corpse was needed for absolute identification. Otherwise, it was just myth and legend. As the great scientists of the time like Owen, Charles Darwin, and Thomas Huxley didn't accept unsupported tales, the tide turned away from belief in fabulous beasts. The phase of "romantic zoology"—animals described by folklore—was over, until the mid–20th century when the field of cryptozoology was officially established.

As we note several times, interest in unusual phenomena of all kinds occurred in postwar time (Regal 2011) and some key figures popularized the idea of undiscovered monsters. The father of cryptozoology is widely agreed to be the late Bernard Heuvelmans, a French/Belgian Doctor of zoology. As a professional, he encouraged amateurs to research descriptions of unusual animals. He took established science to be too stuffy and limiting (Regal 2011) but professed a desire for a high scientific standard in the field. His volume, *On the Track of Unknown Animals* (1955, 1958 in English), was inarguably the most influential cryptozoology book, considered the "Old Testament" of cryptozoology (Sykes 2016). Along with Heuvelmans, cryptozoology was popularized by Ivan Sanderson, a nature writer who studied zoology but never obtained a doctorate degree. Sanderson wrote for popular men's adventure magazines and was a pioneer in nature programing on television (Regal 2011). Heuvelmans and Sanderson worked together at times with the former wishing for scientific credibility for the field and the latter creating splashy and exciting stories. Sanderson went on to publicize other topics considered "unexplained" including UFOs and hauntings, but his scholarship was sloppy (Regal 2011; Naish 2016). A notable example was his failure to identify an outrageous hoax in 1948, declaring that large three-toed footprints found on a Florida shoreline were made by an out-of-place, 15-foot-tall penguin.[2]

Bigfoot is America's most popular cryptid. Before that term was invented, the Yeti ("abominable snowman" or more accurately translated as the "hairy man of the snows"), of Asia, was a pop culture icon (Loxton & Prothero 2013). First brought to Western attention decades earlier, a photograph of a series of alleged Yeti tracks taken by Eric Shipton in 1951 made international news (Buhs 2009). Throughout the 1950s the quest for the Yeti was pursued by explorers, wealthy adventurers, and scientists (Sykes 2016; Loxton & Prothero 2013). A few primatologists and anthropologists were eager to examine the evidence that a relict human-like species still existed and did so with the support of benefactors such as Texas oil millionaire Tom Slick. Investigators were presented with "the Pangboche hand" taken from a monastery in Nepal and a scalp supposedly also from a Yeti, both of which turned out to be of mundane origin (human and known animals) (Regal 2011: 32–53). Sanderson published *Abominable Snowman: Legend Come to Life* in 1961. Sanderson also supported amateur research and had an anti-science streak, feeling that mainstream scientists did not pay enough attention to these claims and real investigators should be independent of universities (Regal 2011: 3).

The 1950s were the Yeti's golden age which likely contributed to the rise of Bigfoot. The North American version of a hairy bipedal ape-man mixed the exciting characteristics of the Yeti with stories of wild men from Native mythology. It is not at all clear if the hairy giants described in myths refer to the same creature we today call Sasquatch or Bigfoot. Teacher and amateur folklorist John Burns collected tall-tale-like stories in British Columbia in the 1920s about what he called "Sasquatch," an anglicized word from the Coast Salish natives describing legendary entities (Loxton & Prothero 2013). The term "Bigfoot" was created in 1958 when a newspaper reported on the footprints attributed to the creature in Bluff Creek, California. The indigenous-sounding word "Sasquatch" was used by some researchers to neutralize the derogatory connotations of "Bigfoot." Both terms merged in the public mind to refer to the cryptid of the Pacific Northwest, but sightings of bipedal hairy ape-like beings eventually were reported from almost every state in the Union (not Hawaii, though give it time). The creatures were of the classic Bigfoot description or of some slightly unique variation, often described with regional monikers such as "swamp

4. Seeking Monsters: Bigfoot and Other Cryptids

ape" (southern U.S.), "skunk ape" (Florida), "Fouke monster" (Arkansas), "Momo" (Missouri), and the "Ohio Grassman."

Bigfoot-like creatures are reported all over the world: the Yowie in Australia, the Almasty in Russia, the Yeti in Asia, and the Yeren in China. Speculation exists among Bigfoot researchers (a few of whom have scientific credentials) that "hairy men" the world over are of separate genus or species than each other and different from that of *Homo sapiens*. They are described, more or less, as *human-like*, creating an ethical question regarding attempts to kill the animal to provide a biological specimen to unequivocally answer the question of its reality. The study of cryptids thought to be related to humans (man-like apes or ape-like men) has been called "hominology"—a field coined in Russia where it was taken seriously in the late 1950s and 60s (Sykes 2016; Gordon 2015).

As Owen stated for sea serpents, any new animal must have a specimen to be formally described. We have no cryptid remains that have been confirmed to be something identifiable as "new." Cryptozoologists suggest reasons why we have no Bigfoot body. For example, they counter that the species is too rare, too smart to be caught by cameras, that the dead bodies are buried or decompose too fast to find. These excuses become more absurd every day as we spot extremely rare animals on trail cameras in remote areas of the world, find remains that can be tested and compared with a known collection of DNA samples, and can even extract DNA traces from secondary sources in the environment such as metal objects, half-eaten food, water, and blood-sucking insects.

P-G Film

The watershed moment for Bigfoot was the Patterson-Gimlin film of 1967, known to Bigfooters as the "P-G" film. The iconic film of a hairy, ape-like creature with breasts striding across a creek bed, and then furtively looking back before continuing purposefully on her way, was well-marketed by the filmmaker, Roger Patterson. Everyone is familiar with the ubiquitous image from frame 352 of the P-G film. This controversial film remains the key piece of Bigfoot evidence, highly disputed to this day as depicting either a genuine Bigfoot or someone in a big suit. The P-G film inspired many to become interested in Bigfoot and cryptozoology in general. About 80% of participants at a Bigfoot conference considered the film a depiction of a genuine Bigfoot (Bader et al. 2010). If it's a man-in-a-suit hoax (which is what scientists and some pro–Bigfoot researchers conclude it is), it's a great one, having lasted many decades without a solid debunking. No one has been able to convincingly reproduce the film clip. In one popular hoax claim, the suit was supposedly worn by Roger Patterson's associate Bob Heironimus and designed by costume maker Philip Morris (Long 2004). This version is disputed by Bigfoot advocates and, most importantly, the suit was never found.

The P-G footage compels many researchers to keep looking for "Patty" (the name given to the celluloid-captured Bigfoot) and her kin. The film has been stabilized, analyzed and criticized countless times. With no new input of information, however, the film is a dead-end as proof of Bigfoot. Like trying to enlarge a grainy photograph, you can't claim details where none were originally recorded. It is amazing how two sets of viewers can look at the same piece of film and either see an obvious hoax or certifiable real animal when it can't be both. The P-G film was brought to the eminent scientists of the time for review. Almost none found it to be compelling as proof of Bigfoot. However, in 1968, Sanderson and Heuvelmans examined a body of what was claimed to be a wild man encased in ice. They found it compelling as evidence and the Smithsonian Institution was interested in obtaining it for study, but eventually the "Minnesota Iceman" was concluded to be a hoax. The various hoaxes of sightings, footprints, and now a faked body had sullied the reputation of Bigfoot and man-like monsters. Derision towards scientists interested in the topic and the overly enthusiastic claims from amateurs soured academics in general on the topic of anomalous primates (Regal 2011).

Through the 1970s, the field of Bigfootery was a mix of passionate amateurs and credentialed scientists—people like anthropologist Grover Krantz and the eccentric Rene Dahinden who gave up everything including his family to find the creature for himself. Krantz and Dahinden often clashed, verbally and physically. As with the ghost-seeking community, money was often at the root of the disputes. Regal (2011) and Buhs (2009) provide detailed chronicles of the several scientists and array of very serious amateurs that examined the evidence for Bigfoot, some of whom dedicated their lives to the search for proof. Both sources reveal the tension between amateurs and professionals and their uncertain roles in the Bigfoot quest. That tension continues today with the increased popularity of cryptozoology as a topic and Bigfoot as America's iconic monster.

Champ

A distant second to Bigfoot's popularity as America's favorite cryptid is Champ, the monster of Lake Champlain, the best-known lake monster in the U.S. (Radford & Nickell 2006). Lake monsters are found all around the world and are generally compared to the iconic Loch Ness monster, Nessie. However, the tale of a monster in Lake Champlain, Vermont, originated long before Nessie surfaced. The first report is from 1808 of some strange creature in the water.[3] "Champ" surfaced in the public eye in 1980 when Sandra Mansi publicized a photograph she had taken three years earlier. This photo was controversial, never verified as evidence of an animal of any kind and later was plausibly recreated as a floating log by Radford and Nickell (2006) in their book *Lake Monster Mysteries*. Lake and local sea monsters continue to be exploited as tourism gimmicks. There remains little convincing scientific evidence of mystery animals in U.S. waters (or anywhere else) but that does not stop the locals from promoting their monsters for visitors. Lake monster enthusiasts will camp out or even live along the lakeshore poised with binoculars keeping a watch for anything unusual in the water. With Lake Champlain being the top location for monster spotting in the U.S., there are those who have established themselves in the media eye as "experts" and formed their own projects, such as Dennis Hall's Champ Quest, to discover proof that there is a genuine unknown animal there. Krulos (2015) states Hall and others are afflicted with "lake monster fever," banking on an American Loch Ness monster. Lake monster hunters equip boats with waterproof audio and video equipment—deep water cameras, sonar scopes, and hydrophones—to capture evidence of an unusual animal. Unfortunately, as with Bigfoot (over)enthusiasts, some will resort to making up "facts" about an animal that has never been located. Hall himself claims to have seen Champ many times. Others obtained underwater sounds they claim are from an animal using echolocation. Amateurs' enthusiasm for Champ's existence looks to some of the public like science-based research.

Crypto-Media

Bigfoot, Yeti and other water and sky monsters were given wide exposure through popular magazines, comics, movies, games, cartoons, and TV shows. As first occurred with UFO enthusiasts, homemade newsletters were written and distributed to a sub-community of Bigfoot enthusiasts beginning in 1969 with the *Bigfoot Bulletin* (Loxton & Prothero 2013). This outlet provided a means for collectors of such accounts to catalog and share news clippings and help to grow the body of sightings data and support that these creatures were widespread. The popularity grew into meetups, conventions, and a cultural explosion in Bigfoot popularity. After the surge in popularity of Bigfoot and monsters in the 1970s (Bigfoot even appeared on the popular television show *Six Million Dollar Man*), cryptozoology was largely relegated to children's entertainment and comedic programming. The TV show *Monster Quest* (2007) was the first series to chronicle rather slap-dash, overly-dramatized investigations of mystery monsters. The show also popularized more obscure creepy cryptids as potentially real such as the Mothman of Point Pleasant, West Virginia, and the Dogman, an American werewolf in Michigan. *Monster Quest* secured a new footing for cryptozoology on television but monster hunters became stars with Animal Planet's ratings winner *Finding Bigfoot* (2011). As with *Ghost Hunters* and its many copycats, *Finding Bigfoot* was a semi-scripted "reality"-type program where a set team of investigators talked to local eyewitnesses then did their own night-time excursions. The promise of the show's title has yet to be fulfilled, but the investigators insist Bigfoot is just of our their reach every time. Critically panned, and a fine example of how *not* to do an investigation, the show is beloved by Bigfoot believers but seen as a joke by active cryptozoological researchers. Perhaps the terrible portrayal of paranormal investigation on TV compels others to attempt to do it better; *Finding Bigfoot*, like *Ghost Hunters*, certainly prompted a wave of individuals and teams to get out "squatchin'" (looking for Sasquatch) all over North America.

As with all fringe topics, cryptozoology was boosted by the Internet with websites and podcasts taking advantage of the media and commercial interest in monster talk. Cryptozoological-themed books, once almost exclusively collections of stories from witnesses and speculative supposition about mysterious creatures, expanded to include scholarly contributions that considered the folklore and sociological aspects of the topic. Books by more skeptical scholars such as Prothero and Loxton, Radford, and Naish were deliberately shunned by the cryptozoological community, eliciting angry tirades and hostilities against the authors.[4]

Your Average Cryptozoologist

The average cryptozoologist in the U.S. is focused on Bigfoot. According to an informal survey by Regal (2011), the mostly male Bigfoot seekers are typically working class, white, between 30 and 40 years old. They frequently identify as naturalists or wildlife enthusiasts. Many are hunters and a few have some biology or wildlife education. Some groups expand to consider other cryptids but Bigfoot is the big prize. Individual researchers specialize in various local cryptids and write books collecting the local lore dug up from newspaper accounts or by interviewing residents. Regal notes (p. 174) that what amateur cryptozoologists lack in education, they make up in enthusiasm, having a strong sense of identity and mission. The amateurs are considered "experts," the "embattled minority" against the "closed-minded" scientists in their "Ivory Tower" (p. 123).

With the decentralization of the UFO/ARIGs and the strongly local ties of ghost/ARIGs, Bigfoot seekers perhaps come the closest these days of all the ARIGs to having any semblance of a unified organization of research. The largest group in the U.S. is the Bigfoot Field Research Organization (BFRO). They maintain a database of sightings, but it has not been systematically studied or published. The BFRO has not so far made an organized effort at analyzing their data set to guide their future research. The head of the BFRO, Matt Moneymaker, is on the *Finding Bigfoot* show. The BFRO labels themselves as "scientific" but Moneymaker himself is not a scientist.[5] While there are rifts between personalities and groups, particularly regarding association with those notables who are accused of hoaxing, most researchers seem to generally get along. At least they are not quite as outwardly antagonistic as the ghost groups are to each other and they sometimes even cooperate and share experiences and data.

Witness accounts are the life blood of cryptozoology; they are the core of the field and the reason for its existence. Cryptids need to be "witnessed into existence" by someone or some group (Loxton & Prothero 2013) which sparks investigators to begin looking for evidence. You will find crypto/ARIGs interviewing witnesses, investigating sighting reports, camping overnight, setting up remote cameras, installing hair or blood traps, making sounds by howling or wood knocking (striking trees with sticks or bats), and listening for replies, preserving casts of what they believe are foot or hand prints, documenting disturbed vegetation or objects, and collecting feces for analysis. Bigfoot-focused ARIGs are driven by a diverse range of ideas. Most consider Bigfoot to be a real flesh and blood animal that leaves traces, has DNA, and could be captured, eventually. Popular candidates suggested for a biological Bigfoot include relict hominins (other human-like beings) from the fossil record. Anthropologists Grover Krantz and Carleton Coon suggested that *Gigantopithecus* (the largest primate so far documented) did not go extinct and may account for Bigfoot/Sasquatch sightings even though what is known about this animal does not match with modern Bigfoot reports. There is no recent evidence that supports the claim that *Gigantopithecus* continued to exist in hidden places.

Bigfooters encounter further difficulty in explaining why the creature is often said to be spotted near human habitation, spying on campsites, and physically harassing the intruders, yet leaves no remains or reliable photo documentation after more than 50 years of searching. Claims by Bigfoot-ARIGs that the creature buries its dead contradict archaeological and paleontological understanding that prompt burial is more likely to enhance preservation. Proposing that a new, large animal is out there yet to be found but not infrequently seen is extremely problematic and unlikely. The need for a genetically diverse population of large animals, and the extensive range required to sustain them, provides ample reason for wildlife biologists to be doubtful of their existence. The record of evidence for Bigfoot and cryptids is plagued with poor scholarship even with a number of scientists involved. The field has not produced scientifically rigorous results.

On the opposite end of the rational spectrum is the paranormal explanation. While psychic powers and UFOs have been associated with Bigfoot for a long while, the growing popularity of out-of-this-world explanations results in a "supernatural creep" away from flesh and blood explanations. Non-natural excuses are proposed in an attempt to explain the extraordinary details in anecdotes such as sudden disappearances of the creatures, their inability to be captured or harmed by gunfire, or connection to UFOs (Gordon 2010). The "Supernatural Bigfoot" idea was promoted heavily by the late Jon-Erik Beckjord who was shunned by "serious" cryptozoologists but was a favorite for outrageous interviews (Buhs 2009; Regal 2011). One ARIG planned to capture a specimen or reputable evidence of what they call the North American wood ape.[6] On the opposite extreme are those who assume Bigfoot and other mystery monsters are inter-dimensional or have psychic- or super-powers found in no other lifeforms and, thus, can elude human detection. People with no stake in acceptance by an academic or government establishment are more likely to take this paranormalist approach. Some revel in an unconventional persona, blurring boundaries between the cryptozoological and occult communities (Bader et al. 2010). As with the UFO and ghost communities there are Bigfoot "contactees" that claim to have a deep personal, even telepathic, connection to local populations of Bigfoots they claim live nearby. Others compare such claimed experiences to a religious state (Sykes 2016; Buhs 2009).

Bigfoot captures most cryptozoological attention but ARIGs also investigate reports of lake monsters, out-of-place big cats such as pumas or black panthers, the Jersey Devil, chupacabras (a variant of which are called "Texas blue dogs"—popularized by an episode of *Monster Quest*[7]), and other unusual animal sightings. In some lucky cases, a body (albeit a strange one) has been discovered and photographed with the resulting story making a media splash. Witnesses and cryptid enthusiasts will invariably speculate it *could* be a new species but scientific investigation identified these cases as individuals of known species. Media stylized "mystery creatures" turn out to be an unfortunate native animal with a disease or uncommon genetic condition, or an imported out-of-place species. Carcasses of domestic or native animals can look "weird" and may not be not immediately identifiable due to effects from decomposition or environmental exposure that the average person is unfamiliar with. Such mysteries are quickly labeled "monsters."

Professional Cryptozoology

In 1982, a group of scientists interested in the search for rumored animals formed the International Society of Cryptozoology, producing their own journal and newsletter. Their goal was to provide a unified effort to pursue the field and to provide goals and a system for sharing information. Many scientists from around the world joined the ISC (Regal 2011). Cryptozoology was on the path to respectability. Those who subscribed to fringe explanations for cryptids (aliens, alien pets, interdimensional beings, spirits, thoughtforms, etc.) were relegated to the fringes. However, most Bigfoot hunters didn't like the ISC which floundered and eventually disappeared due to mismanagement in 1998 leaving the field without an official society, no overall guidance, outlook, or publication.

Loren Coleman, an author with a background in social work, considers himself a full-time cryptozoologist. Coleman founded and maintains the International Cryptozoology Museum in Maine and has resurrected an official organization, called the International Cryptozoology Society (ICS). Zoologist Dr. Karl Shuker also began publishing the *Journal of Cryptozoology* in 2012, but most of today's cryptozoology media runs amok on the Internet and television in an insidious blend of fact and fiction without discernment. The subject of monsters, real or legendary, appears to be growing in popularity, most notably online. As with other paranormal investigations, credentialed scientists are rarely in the mix (Loxton & Prothero 2013). Within academic anthropology, the pattern has been to have one Bigfoot-friendly professor at a time. For much of the later twentieth century, this was Dr. Grover Krantz (Regal 2011). Today, the current pro–Bigfoot scientist is Dr. Jeff Meldrum, an anatomy and anthropology professor at Idaho State University and regular headliner of cryptozoology-themed TV shows, podcasts, and conferences. Geneticists, such as Dr. Todd Disotell of New York University, have stepped up to assess supposed DNA evidence claimed to be from man-like apes. Professor Emeritus from Oxford University, Dr. Bryan Sykes, analyzed DNA samples from around the world hoping to find a unique signature. His work did not locate the enigmatic Bigfoot or a Yeti as hoped but his book revealed that an academic reputation, connections, and funding provided access to tools and experts ARIGs wouldn't have.

For example, he noted that evidence carefully collected by Bigfoot researchers was treated shabbily by labs because of the assumption that it was nonsense or that they would not get paid (Sykes 2016). The Olympic Project was another attempt at an organized approach to Bigfoot research described as an "association of dedicated researchers, investigators, biologists and trackers committed to documenting the existence of Sasquatch through science and education."[8] So far, no such project or test result has provided the goods to state Bigfoot is real.

There is no academic course of study in cryptozoology or no university degree program that will bestow the title of "cryptozoologist." Some skeptical-minded researchers are steering the field more towards scholarship in folklore, history and social sciences, and away from capturing a flesh and blood "cryptid" that some enthusiasts hope will overturn scientific knowledge. After decades of intensive search, we still have no Bigfoot or lake monster to show for such efforts. One common argument made by cryptozoologists is that indigenous people have special knowledge of local animals, including cryptids. Or, those who hunt or fish regularly can more readily recognize animals in the wild. Cryptozoologists frequently tout the findings of European explorers who brought strange animals known to locals into the scientific domain. The era that brought us the discovery of the giant panda, mountain gorilla, and okapi largely came to end with the last stages of European colonialism and these animals are now well-known from museum specimens and zoos. The giant squid served as an example to cryptozoologists that legendary monsters may indeed be alive in the sea. While we now have bodies and several verified and recorded live sightings of the giant squid, wariness of ambiguous testimony as evidence dates from the very early days of science (Lyons 2009). There is good reason for that doubtfulness. Eyewitnesses, no matter how experienced, are prone to perceptual distortion in poor conditions and bias to see what they think they should be seeing. Ideally, anecdotes can be useful in that they *should* lead to confirmation of the observations and better evidence. This does not happen with modern cryptids. Modern new species are either visually similar to known species or a complete surprise unrelated to mystery reports of strange animals. Many new species are found in museum drawers, collected decades ago and never fully examined.

In a unique twist, Biblical Creationists, who deny the process of evolution and the geologic evidence for a multi-billion-year-old earth, comprise a subset of cryptozoology advocates. They are well-funded and able to conduct expeditions with a goal of finding a living dinosaur that they think would invalidate evolution. It wouldn't. Yet seekers of living dinosaurs and other cryptids are motivated by the Biblical literalist view of a divine cre-

4. Seeking Monsters: Bigfoot and Other Cryptids

ation and to show that conventional science is wrong (Loxton & Prothero 2013).[9]

Bigfoot has become synonymous with "hoax" (Buhs 2009). This association, which also tarnishes the investigation of ghosts and UFOs to a lesser degree, creates an embarrassing situation for serious-minded seekers. Several Bigfooters carry a reputation as known scammers, the most notable being Rick Dyer who attempted to capitalize on several concocted shams starting with the Georgia Bigfoot body of 2008 (Loxton & Prothero 2013). This sensational story was brought directly to the public via press conference. Once images of the dead "Bigfoot" were circulated, it was quickly exposed as a costume with animal innards added for effect. Dyer's scam followed a tradition of Bigfoot hoaxing that included the Minnesota Iceman that fooled Heuvelmans and Sanderson. But Dyer wasn't done jerking the chain of Bigfoot believers. After the Bigfoot-in-a-freezer plan failed, he constructed a model named "Hank," concocted a story that he shot it, and took it on a pay-per-view tour. He failed to make significant money off these gaffes, but drew many hopeful Bigfoot hunters in to his side-show. Other individuals with highly questionable activities include Tom Biscardi who runs the full-time Bigfoot-searching company called *Searching for Bigfoot, Inc.* Biscardi is most famous for being the spokesperson who promoted the finding of Dyer's Georgia Bigfoot as a real animal carcass, and conducting the press conference that was covered by CNN. He later stated he'd been hoaxed himself, fooled by a costume and animal entrails. Todd Standing of the *Sylvanic* project that began in 2005, presented close-up photos of Bigfoot faces framed by tree branches in 2011. While charming and cute, the faces looked like puppets and did not resemble living creatures.[10] *Finding Bigfoot* cast member Cliff Barackman was not at all impressed during an episode featuring the Sylvanic site.[11] Standing had also gained attention by appearing on the popular show *Survivorman: Bigfoot*. This parade of rejected evidence framed cryptozoology as bunkum, taking your money and giving you an ultimately disappointing reveal.[12]

A few other Bigfoot projects were funded by private sources and crowd-sourced donations such as those by veterinarian Melba Ketchum to test alleged Bigfoot DNA samples (Ketchum et al. 2013), the Erickson project to capture photos of creatures in their habitat, and the Falcon project to fund a blimp to search for heat signatures from large animals in the woods. None of these projects have produced compelling evidence for cryptids. Ketchum collaborated with the Erickson project to produce a 2013 paper which was ridiculed for unscientific and absurd claims (Hill 2013). Poor scholarship from those who claim to be professional has left the field in disrepute.

Crypto-ARIGs assume there are mystery animals out there to identify and find. Many are passionate and sincere in their belief that the mystery animal exists. (See Sykes 2016.) As such, they give deference to every report of a sighting, often without critical questioning. As with the ghost seekers, cryptozoologists are convinced they will be the ones to solve the mystery and make history. With the lure of mystery and money undermining diligent and ethical research, the field of cryptozoology has serious credibility problems.

5

UFO Spotters

Flying Saucers

Millennia of historical records exist of strange things people reported seeing in the sky. These reports are reflective of the cultural beliefs of the time, including the "mystery airships" at the end of the 19th century. Only recently, at the dawn of the Space Age, has the phenomena become associated with extraterrestrial beings. Saucers began flying in earnest in 1947 and within weeks 90 percent of Americans were familiar with the term (Thurs 2007). The sightings integrated into a new mythology that included prophets, mystics, seekers, and hoaxers (Moseley & Pflock 2002). A darker ufology (study of unidentified flying objects, UFOs) erupted in the second half of the 20th century, when a wild array of speculation arose about government conspiracies, alien technology, abduction of humans, cattle mutilation, hybrid experiments, and ancient alien gods, beginning in the U.S. and spreading abroad (Eghigian 2014). Interest in space visitors is fueled by our fear and fascination with what is beyond earth and beyond human technology and intelligence.

Investigating UFOs

The military, particularly the newly formed U.S. Air Force (USAF), were the first to be interested in UFOs, studying them as a national security risk, perhaps an enemy device (Sheaffer 1986). A scientific approach was desired, with reputable astronomers, astrophysicists, and other scientists involved in the issue.

The USAF undertook their first investigation, Project Sign, from 1948 to 1949. The results were less than exciting, showing there was nothing to fear. Then Project Grudge later that year was a reorganization of the investigation. The public saw these as "Project Saucer" in response to the growing number of reports of the objects (Thurs 2007). Project Blue Book, begun in 1952, was another reorganization of the investigation but by now the work was done by a much smaller number of personnel (Thurs 2007). These reports standardized the collection and systematically analyzed claims of UFO sightings. The Robertson Panel issued a report in 1953. All these government-sanctioned investigations concluded that there was no evidence of a direct threat to national security or risk to the U.S. from these objects, a finding that would also be replicated by the UK Ministry of Defense office. While many sightings of strange objects remained *unexplained* (hence the name), that did not equate to the sightings being *unexplainable*. Residual cases allowed some people to conclude there was something genuinely mysterious going on, but there was not enough information to conclude what, if anything, was responsible for the reports. Ultimately, the conclusion about UFOs was that people were making mistakes, lying, or hoaxing—a conclusion that did not sit well with those who were convinced they saw something otherworldly. The public was not entirely assured that these investigations were anything more than public relations efforts.

The U.S. government's secrecy and habit of providing conflicting information to the media fed a growing conspiracy sentiment that there was something to hide. Even though the Air Force ceased official study of UFOs, their exit was not graceful since unknown aerial phenomena continued to be reported. The Robertson panel had recommended an educational effort to "debunk" this topic, a plan that would hopefully reduce the distracting public interest and a potential danger for panic in society. They hoped a two-year effort would be enough. Obviously, it wasn't.[1] Twenty-two years after the 1947 explosion of national interest regarding flying saucers, no scientific investigation was deemed adequate to all parties. Proponents felt that scientists refused to recognize the scope and importance of UFO reports (McDonald 1972). Instead of tamping down the interest, rejection of the phenomenon by official government agencies and the scientific community bolstered the growth of independent civilian groups formed to investigate flying saucers. They adopted official- or military-sounding names. Ufology became a field (like cryptozoology and ghost hunting) where a layperson without any formal post-secondary education could become an "expert" (Denzler 2003). The first group appeared in 1952—the International Flying Saucer Bureau. Led by Albert Bender out of Connecticut, the IFSB published a newsletter with several hundred subscribers (Mosely & Pflock 2002). 1952 was the start of what Thurs (2007) called "sauceritis." The Aerial Phenomenon Research Organization (APRO), also founded in 1952 by Coral and James Lorenzen, eventually received support from PhD scientists. Civilian Saucer Intelligence began in 1954. BUFORA in the U.K. started a 24-hour

5. UFO Spotters

hotline to accept reports and attempted to use rapid response teams to investigate, a model later UFO/ARIGs would follow. Since 1974, NUFORC (National UFO Research Center) has maintained a toll-free hotline for reporting UFO incidents. According to their website,[2] various governmental agencies refer callers to this organization for logging into their database.

Groups, large and small, local and national, attempted a systematic approach to collecting and investigating UFO reports (Eghigian 2015). The most prominent of the civilian groups was the National Investigations Committee of Aerial Phenomena (NICAP) under the direction of Major Donald Keyhoe, formerly of the U.S. Marine Corps with involvement from former officials from the military and intelligence communities. NICAP took a stance that aliens had increased observation of earth after the atomic bomb detonations and the USAF was covering up real information on UFOs. This idea became common throughout the investigator community (Sheaffer 1986; Thurs 2007). Keyhoe and NICAP pushed for congressional hearings into UFOs. As with other ARIGs, early UFO groups felt that the official sources were overlooking critical details and they were determined to seek out "the truth" for themselves.

The height of public and political interest in UFOs came after a wave of sightings reported in 1965–66. With increasing pressure for investigation, the USAF funded an investigation by the University of Colorado to stand as the final conclusion on the matter. Dr. Edward U. Condon, was the Scientific Director of the 1968 independent committee that was tasked to see if there was any scientific knowledge that could be gleaned from specific UFO cases. The "Condon Report" concluded that no scientific value was gained from UFO claims. The report, freely available and made into a mass market paperback, was widely criticized by UFO proponents and was the end of both Project Blue Book and serious scientific interest in UFOs (Eghigian 2015; Denzler 2003). Instead of settling the issue of UFOs, however, it created a new arena for questions (Thurs 2007). A few scientific people remained interested but UFO groups eventually became outwardly antagonistic to the science community. Press interest in UFOs turned from serious to silly. Ufology devolved into personality cults, pseudo-religious, and New Age collectives, producing no content worth citing or taking seriously (Moseley & Pflock 2002). Facing other typical organizational pressures, many civilian UFO groups folded.

MUFON

MUFON, the Mutual UFO Network, emerged from the Condon report aftermath in 1969 and today is the largest remaining UFO-ARIG with branches that cover every state with appointed directors for each region. Participants in the state or regional groups pay dues to the national overarching organization but the investigators are local. MUFON field investigators assisted CUFOS, the Center for UFO Studies, a group founded in 1975 by former Project Blue Book consultant, Dr. J. Allen Hynek. CUFOS consisted of scientists and professionals as consultants and provided grants for UFO researchers. CUFOS still maintains an archive of material and their UFOCAT database.[3] CUFOS became the J. Allen Hynek Center for UFO Studies upon Hynek's death in 1986.

MUFON remains the main investigative network in the U.S. They provide a training manual that can be purchased by anyone. To become a MUFON investigator, one must pass an exam based on this manual. Two former MUFON members told me the exam was not difficult and did not require anything more than studying the MUFON manual's contents. No other credentials are needed. MUFON has attempted a standardized training protocol, which is more than can be said for other paranormal investigation fields. The individual branches operate somewhat independently with little oversight so it's unclear how "organized" the network actually is. There have been problems with the MUFON leadership in the past years as some branches and even the main headquarters have drifted into emphasis on areas' claims of abduction, disclosure, and "exopolitics" and away

A MUFON convention audience in Bucks County, Pennsylvania, 2016. Photograph by Kenny Biddle.

from a scientific tone about what people report in the sky.[4] They invite speakers to their conferences that some within the UFO community feel are not credible or are blatant publicity seekers. There has been a call from the less conspiratorial members for the organization to pull back from discredited subject areas and return to "feet on the ground" immediate investigation of reported sightings—a formidable task.

Many curious scientists and professional people investigated ufology but left the scene when they discovered there was little to no merit or substance to it. Though J. Allen Hynek is considered the founder of ufology, other respected people involved in the discussion of what to do about these flying objects included computer scientist Jacques Vallee and skeptical astronomers Donald Menzel and Carl Sagan. Sagan would become a powerful force for skepticism regarding UFOs but in the 60s, he was genuinely interested in the idea of intelligent aliens. In 1969, a scientific symposium was held under the auspices of the American Association for the Advancement of Science (AAAS), the publishers of *Science* journal. Dr. Sagan spearheaded the efforts (Denzler 2003) but many other scientists, including Condon, opposed it as a waste of time and effort (Sullivan 1972). According to Eghigian (2015), the boundary remained porous as some scientists and other professionals insisted UFOs warranted scientific research. After the Condon Report, the focus of any academic UFO/alien interest was in belief, public attitudes, memory, folklore, psychology, and conspiracy ideas. Rejected by government officials and scientists who thought they might be crazy, those who experienced UFOs or claimed contact with real aliens distrusted the establishment. There was a self-perpetuating isolation of UFO believers from the scientific establishment. Attempts to make sense of the phenomena as cultural or psychological ended up reinforcing distrust.

By the 1970s, with the USAF out of the loop, sightings were being reported directly to UFO/ARIGs and the mass media. The field of ufology deliberately bypassed institutions and connected directly to the public. They considered themselves serious researchers focused on a serious problem for humanity. A few groups collected detailed sighting reports for statistical analysis but most focused on individual cases. They drew upon what they thought were scientific methods. They formed institutions and journals that sounded legitimately scientific. But without the unifying sparring partner of the Air Force, UFO and alien topics blended into the expanding paranormal media culture of the 1970s. In response to the growing media focus on fringe topics like UFOs, a group of philosophers and scientists formed the Committee for the Scientific Investigation of Claims of the Paranormal (CSICOP) in 1976. The CSICOP name was punned by the paranormal community as "psi-cop," the scientific police and enforcement agency.

In the 2000s CSICOP was renamed the Committee for Skeptical Inquiry (CSI), causing confusion with the "crime scene investigation" acronym used for a popular television franchise.

Contactees and Abductees

A diversion from the path of investigating reported sightings occurred with the "contactee" movement that grew in part out of the occult underground of mid-century southern California. This fringe community blended science fiction-like themes with occult practices such as extraterrestrials appearing through astrally-channeled messages, tales of advanced technology, spiritualist mediums, and science jargon. The aliens were "space brothers" warning us of impending atomic or environmental doom. The most prominent contactee was "Professor" George Adamski who called himself a "scientist" or "one who knows" (Thurs 2007). The Amalgamated Saucer Clubs of America was a contactee-friendly outlet to report encounters (Thurs 2007). Adamski's book in 1953 (written with Desmond Leslie) along with Keyhoe's book in 1950, heralded the era of amateur ufology (Eghigian 2014). The religious connotations of the contactee movement created a vast distance between the subject and scientific study (Denzler 2003) though psychological and sociological research of contactees and UFO religions was important to the formation of the concept of "cognitive dissonance."[5]

Less spiritual reports of alien beings or "close encounters of the third kind" as dubbed by J. Allen Hynek, began to emerge in the mid-1950s. A wide variety of strange creatures were reported in association with UFOs. Vallee compiled many of these cases and drew parallels to earlier tales of fairies, elves, and other magical people of folklore. Vallee concluded that perhaps the historical reports could have been extraterrestrials. The relatively harmless alien contact turned sinister as the occupants of the UFOs reportedly were getting up close and very personal with Earthlings. The key case that made media headlines was that of the abduction of Betty and Barney Hill in 1961, later made into a book and a television movie. Reports of abductions proliferated after 1975 (Sheaffer 1986). The UFO abduction subculture exploded in the 1980s thanks in part to Travis Walton who became a media sensation by claiming, just weeks after the airing of the Hill abduction movie, to have been taken by a space ship. At the foundation of the new abduction movement was the use of hypnotic regression. Considered legitimate in the 1960s, hypnosis to enhance memory of a traumatic event was used in the Hill case which served as a model for subsequent abduction research. Artist Budd Hopkins is credited with rocketing the abduction idea into the mainstream and promoting the use of hypnosis in abduction claims. Hypnotic regression came under severe scrutiny during the Satanic Panic era of the 1970s and 80s when people reported demonstrably false abuse claims related to devil worship. The technique was discredited but remained in use within alien abduction research.

In 1981, an organization called the *UFO Contact Center International* was formed to help people who had these bizarre experiences. Science fiction and horror novelist Whitley Strieber produced books he said were accounts of his own amazing abduction experiences. Psychologist John Mack of Harvard University and historian David Jacobs of Temple University made academic waves with their conclusions that these events were happening as people described. Mack saw abduction as evidence of a realm outside material reality while Jacobs promoted the idea of sinister alien-human hybridization. The issue of UFOs as carrying occupants from outer space and their associated involvement with humans further divided the now even more colorful field while legitimacy remained elusive (Thurs 2007).

Today's Ufologists

Stan Gordon is an electronics technician and radio communication specialist who can rightly lay claim to the top spot as the most diligent UFO (and anomalies) investigator in the country. Gordon has been investigating UFO reports since 1965 and still does today. In 1969, Gordon set up a UFO hotline phone number for locals in western Pennsylvania to call (24 hours) to report a mysterious incident. The local and state police as well as the local planetarium and others who receive reports of strange lights or weird animals, direct claimants to Gordon. He also documented the addition of paranormal experiences to UFO sightings as people reported psychic experiences, otherworldly events, and encounters with hairy Bigfoot-like creatures during the 1973–74 "flap" of UFO and paranormal reports in Pennsylvania (Gordon 2010). Today's determined UFO investigators, like Stan Gordon, wait for a call about a sighting then attempt to reach the location as soon as possible to talk to the witness. They always have their cameras ready. Space debris, meteors, or rocket launches are often reported as UFOs. Surprised eyewitnesses report details about the sighting including information that is later found to be untrue or exaggerated. Faced with a mundane explanation, they will deny that the incident can be explained other than by processes "not of this world" (Sheaffer 2016).

UFO/ARIGs face the most difficult conditions for data collection. UFOs are fleeting, difficult to record and describe due to their location in the sky, are potentially very far away, and often reported at night. UFOs are reported to travel across states. The sky is ever more cluttered with man-made flying objects like weather balloons, other experimental balloons, Mylar and helium party balloons, Chinese (sky) lanterns, and remote-controlled flying objects (including some that are modeled after the stereotypical flying saucer). Further complicating investigation of UFO sightings are civilian observations of secret military experiments that officials did not want to be public. In remote western U.S. desert areas, serious UFO spotters looking for anomalies occasionally spotted secret tests of new technology. Such scenarios introduced a unique level of intrigue and danger to this ARIG category not commonly seen in the other categories.

With everyone carrying a mobile camera for still and video shots these days, the ubiquity of surveillance cameras, and the many technically sophisticated eyes on the skies, the fact that no good UFO evidence has been authenticated caused a decline in interest in nuts and bolts UFOs and a rise in dramatic ideas with a foundation in impossible physics and science-fiction. The spectrum of UFO researchers is wide. Rifts exists between those "nuts and bolts" investigators of aerial phenomena and those invested in a conspiratorial view that the government is hiding data and hushing up the "truth." This line of thinking became a slippery slope that led to bizarre scenarios of Men in Black silencing witnesses, brainwashing, and the idea that "Reptoid" aliens are in control of world government—an idea made popular by former UK athlete and sports broadcaster, now author and public personality, David Icke. Notably, these ideas are rooted in pulp fiction.

Specialty newsletters and journals were important sources of info prior to the Internet. Modern amateur UFO research began long before the days of the Internet. Groups communicated via mailed newsletters and flyers physically posted on bulletin boards. You can access thousands of international websites and Internet forums dedicated to news and views based on the belief that the "truth" is out there—aliens are visiting us and the government is covering it all up. There isn't a moment that you can't find a UFO-themed show on cable television, especially on those billed as "educational" channels. Countless videos are uploaded to YouTube with opinions and presentations on the entire gamut of UFO and alien claims. Information for enthusiasts is overwhelming in volume in dozens of languages. State or regional MUFON conferences remain well-attended with the biggest gatherings, the International UFO Congress, attracting about 1,500 attendees (Krulos 2015) and the Contact in the Desert convention which may reach 2,000.[6]

Crashed Saucers and the Modern Era

Crashed saucer stories were an early staple of UFO reporting but a series of crash hoaxes revealed in the 1950s made the topic verboten within ufology for decades. The collapse of the civilian UFO investigation bureaus and the rise of UFO conspiracies in the 1970s re-energized the topic. The world famous "crash at Roswell" became the hallmark UFO story. Buried and forgotten soon after the publicity circus of 1947, the resurrection of Roswell, New Mexico as an alien ship crash site began in the late 1970s due to researcher Stanton Friedman. Soon others followed. By the 1980s, crash research was becoming the primary focus of "nuts and bolts" researchers. In 1994, Roswell was rediscovered by America popular culture making the town a huge tourist draw. Combined with other narratives, the modern UFO milieu includes Men in Black, secret Area 51 and the legendary Hangar 18. UFO cults like the Raelians and the ill-fated Heaven's Gate bloomed. As sociologist Brenda Denzler (2003) says "there was no reining it in now" (p. 31) as the UFO theme was international and mainstream. UFOs experienced this resurgence in interest thanks to the capability to participate in online forums and share stories, photos and videos and opinions worldwide. Photos and videos promoted as evidence to support the strange reality of UFOs find their way to every corner of the world via mystery-mongering and conspiracy websites dedicated to mainstreaming fringe causes for UFOs. A research-based organization and media company called *Open Minds* presents its mission to investigate and report evidence of extra-terrestrial, UFO, and other phenomena to a global audience.[7] They produce a website, magazine, radio show, and video programming with specialty reporters to write, distribute, and promote this view of ufology (Krulos 2015). Online, speculation rules; the skeptical viewpoint is rejected. Any critical thinker who provides evidence of a hoax photo or video can expect to be labeled a "disinformation agent"—that is, a person who is deliberately spreading lies to throw seekers off the real truth.

Yet, as with all claims of mysterious phenomenon, hoaxes abound. From hubcaps thrown through the air to a man-made saucer suspended from fishing line, faked UFO photos have always been part of this subculture. Today, getting a photo or video to "go viral" on the web and generate buzz often seems to be a goal. Those just a little bit skilled in photo manipulation software or computer graphical interface animation can produce hoaxes that will get Internet mileage. The jig is typically up in short order when debunkers pick it apart. The mysterious photograph with a fun story is still regular fodder for local news and social media. While MUFON representatives are still asked to comment on UFO sightings in the media, the

78   Scientifical Americans

research and investigation of such claims in the real world is difficult and unglamorous.

Decades ago, grainy NASA photos were interpreted as a "face on Mars" by mystery-mongering commentators. Today, the technology-based, armchair researcher is provided with ample anomaly fodder due to the avail-

A decent UFO photograph can be created by using household items or equipment parts such as this cap used to hold a spare tire. The size of the object is irrelevant as scale and distance aspects are lost in the resulting photograph. Photographs by Kenny Biddle.

ability of high-resolution photos from NASA's Mars Curiosity Rover. Those with an abundance of free time and motivation meticulously scan the robot's photos searching for any rock, shadow, or glitch that looks suspiciously like bones, animals, or some remnant of life. They have spotted interesting mineral formations, weathering features and marks left by the robot itself but no alien selfie has emerged. Major news outlets pick up these stories thanks to social media and they become what passes for news in the 21st century. No journalism is necessary. Debunkers also have handy investigatory tools—the Internet can provide a researcher with maps of the land and sky, exact timing and pathways of civilian aircraft, photos, and info about military prototypes being tested, weather conditions, and meteor or space debris confirmations.

Website revenue from ads, book sales, personal appearances, and television spots allows some to make very good money from the UFO industry. But these riches pale in comparison to those made by media production companies who have mined the UFO mythology.

Alien TV

Ufology was brought directly to the public via books, self-published journals, magazines and conferences, TV, and movies. Radio provided the earliest boost to flying saucer popularity to be followed by stories in men's magazines like *True, Fate,* and *Amazing Stories* which coincided with the science popularization in the 1950s. Discussion about the portrayal of aliens in movies would be a whole other book but was, undoubtedly, the most powerful influence on people's perception of the reality of aliens and UFOs as extraterrestrial craft—how they looked and behaved. The contactee movement was influenced by the movie *The Day the Earth Stood Still.* The rise of the big-eyed, spindly-appendaged, gray alien followed in the wake of television depictions and Spielberg's *Close Encounters of the Third Kind.* The "grays," as with the hairy ape-man Bigfoot and the transparent floating ghost, are inescapable pop cultural media icons.

Interest in UFOs and aliens swelled thanks to television. TV shows like *In Search Of...* (1976–1982), *Unsolved Mysteries* (1987–2002), *Project U.F.O.* (fiction based on Project Blue Book, 1978–79) and *The X-Files* (fiction, 1993–2002, 2016) were instrumental in sowing interest in mystery about UFOs, Men in Black, alien visitation, and abductions of people and vehicles.

A related concept of pseudo-archaeology postulated that alien civilizations visiting Earth must have had a hand in building the pyramids and other ancient wonders. Authors like Erich von Daniken who promoted extraterrestrial influence on human history were and still are tremendously popular with the public. The History Channel took advantage of this interest with the show *Ancient Aliens* that used nonscientists as professional-looking talking heads spouting speculative fiction that sounded convincing to those who knew nothing of ancient history. Pseudo-archaeology is a growing platform for non-experts. Several flavors of Biblical literalism and New Age pseudohistories appear in movies, books making inroads with the public ideas about human history.

The TV show *Chasing UFOs*, which aired on the National Geographic channel in 2012, attempted to capitalize on the popularity of the ghost- and Bigfoot-chaser programs by constructing a purposefully diverse team of ufologists. The show failed and was even criticized by its stars before it was cancelled. It was not compelling television to wait for things to appear in the sky or to chase after something that does. As frequently occurs with media portrayal of investigators, the individuals had stereotypical and exaggerated roles—the tech person, the token female, etc. Drama is heightened (and manufactured). Paranormal reality TV shows are designed as entertainment for the curious, not science documentaries to discover truths. If the tempo is too slow it will be sped up by giving "reality" a boost.

6

Twenty-first Century ARIGs

Paranormal Technology and Gadgets

Use of technology is a common thread through ARIG subject areas. ARIGs will state that better technology is needed to help them pin down definitive evidence. Technology and science remain inextricably linked in the public eye. The historical use of gadgets and electronics in ARIGs' fields, particularly ghost hunting, came very early in the modern age (related to the invisible propagation of electricity and radio waves), and was enhanced by the Atomic and Space Ages when science was directly associated with impressive and advanced machinery. In the public eye, scientific observation is inseparable from use of technical equipment. Devices promote an image of objectivity (Thurs 2007). There seems to be an unstated importance in showing elaborate apparatus as part of portraying "science" (Beveridge 1957; Toumey 1996).

In the 21st century, impressively high-tech equipment is within the price reach of many amateurs. Today's Bigfooter has a FLIR camera and the UFO hunters can get access to databases over their smart phones. GPS provides precise location information. Digital cameras can be installed all over a house as well as on game trails. Today's ARIGs even use motion-sensing video game controllers and remote-control drones and robots. Mainstream science rejects the premise that this investment in equipment advances knowledge about paranormal topics.

The iconic ghost hunter has gadgets with which to locate spirit energy or communicate with the dead. In attempting to answer where this gadget affinity came from, I found suggestions ranging from Harry Price in the early 1900s to *Ghostbusters* in the late 1900s to *Ghost Hunters* in the 2000s. In fact, communication with spirits was a theme that arose almost immediately upon invention of the telegraph, photographic camera, the radio, and the phonograph. Ghostly activity adapted to modernity and can be captured by telephone, answering machines, mobile phones, televisions,

A typical carrying case of ghost-hunting equipment. Items include cameras, EMF meters, infrared temperature meters, digital audio recorders, and motion sensors. Photograph by Kenny Biddle.

and audio and video recording devices, both analog and digital. The key is electricity. Jeffrey Sconce's *Haunted Media* (2000) walks the reader through the relationship between electronic media and American culture from the beginning of social awareness about electricity, to reaction to the telegraph, radio and television. Sconce says (p. 202) "electronic media have always indulged the fantasy of discorporation and the hope that the human soul, consciousness, or subject could exist independently of his or her material frame."

The seeds of modern ghost hunting gadgets (like EMF meters and electronic transcommunication devices) were obvious in the 1850s. Ghost/ARIGs of today are drawn to electronics the same as the spiritualists and psychical researchers at the beginning of the modern age were to gadgets and apparatus. Early on, electricity was believed to be a mystical and divining substance that animated the body and soul (Sconce 2000). The telegraph and radio sent a signal without a body to deliver it. They transcended space and cut down time. Perhaps, many speculated, humans could create devices to reach the spirits of the dead, those "beyond" time and space. It's hard to imagine that at the time (1844), the telegraph was astonishing to the public. How could people so far apart speak as if together? Four years later, the Fox Sisters could communicate via knocks to the spirits. Sconces says this link between Spiritualism and the new technology was no coincidence in time. People believed the dead spirits were not gone but out there somewhere. The "ether" was alive with spirits or messages from others and we might finally devise the gizmo that could capture them. Spirit voices perhaps moved through the ether just like electricity and radio waves. Thomas Edison attempted to invent an apparatus that would capture transmissions and allow the living to hear voices of the dead. Though he was America's top technology inventor, the telephone to the dead was not forthcoming. Curiously, it was also common thinking at the time when radio or "wireless" became popular that similar broadcasts from beings from outside Earth could be captured (Sconce 2000). As technology enabled humans to reach across vast distances, there seemed a reasonable possibility discussed at the time that we could reach those that had passed to some other plane of existence. The promise that gadgets can prove life beyond death continues to the present time.

Tinkering with and mastering gadgetry comes naturally to many. It has been suggested that some enjoy technology-based activities as symbolic compensation for a lack of power or mastery in other areas (emotional and social) of their lives (Kleif & Faulkner 2003). Ghost/ARIGs tend to lean heavily on their equipment, embellishing and promoting their "techknowledgy" (Sconce 2000) which suggests that this compensation is occurring in some ghost seekers, but not all as more paranormal investigators have given up on finding spirits through ghost tech gadgets. "The use of tools in the field ... has seemingly clouded the minds of those who are attempting to verify anomalous phenomena through them. Many people think that the mere use of these tools is science and having anomalous readings with them serves as evidence of the paranormal."[1]

Gadgetry related to ghost hunting was greatly enhanced by Harry Price, as described in Chapter 2. But the full uniform of ghost hunters with their array of electronics reached its apex with *Ghostbusters* (1984) where frustrated scientists achieved fame and scholarly prominence by saving New York City from an infestation of psychic energy. ARIGs may not deliberately be imitating characters in a fictional movie but the comparisons between real life ghost hunters and fictional ones do suggest the movie has been influential.

ARIGs sometimes lament the cost of expensive equipment because, they note, they have no funding sources like established scientific programs do. Some do attempt to gain their own television shows, win contests or solicit private investments or sponsorships to account for some funding. Such income will be rare. But the need for having equipment, especially the most modern versions, seems deeply embedded with the idea of any paranormal investigating. Reliance on equipment by ARIGs and the obsessive need for some ARIG participants to have the latest and greatest technology has created a niche paranormal investigation market directed towards ARIGs. Advertisements for these companies are sometimes found on the ARIG web sites. There are "ghost hunter" stores selling specialized gadgets as seen on the TV shows. At least one, *Ghost Stop*, branched out to include a *UFO Stop* store. While cryptid hunters typically utilize little in the way of special equipment beyond camping and photographic gear, there are novelty products such as Bigfoot callers.

Advocates of devices and sciencey equipment consider the apparatus as objective and a "faithful register" of facts (Harvey 2013). Eyewitness accounts or artists' renditions, such as those of the ubiquitous sea serpents, were not enough to authenticate the phenomena. If they can't get physical proof, some sort of image, video, sound, or environmental signal is desired. However, 21st century technology is astounding—we can detect subatomic particles, planets outside our solar system, and decode DNA—but our technological capabilities have not been successful in confirming consciousness after death, haunting phenomena, strange lights in the sky or mysterious animals that have remained hidden to our eyes.

ARIGs on the Web

The idea of community was changed because of the Internet. The lowered cost and greater availability of computers and Internet connectivity in the 21st century allowed people sitting at home to instantly access information and connect with those of similar interests and ideas. A megaportion of the world's knowledge is at our fingertips and we can speak to others across the globe in real time. This amazing technology was essential for the explosion of interest in the paranormal (and, likewise, skepticism and criticism of paranormal claims). It is near-impossible to imagine life before the Internet. How did these niche interest groups communicate and stay organized pre–Web? They had homemade newsletters, journals (usually a low-budget affair) and written communication. Flyers were hung up at libraries, grocery stores or community centers. Sweeping changes brought by the Internet led directly to the ARIGs of modern times. Paranormal culture effectively utilized the web from the beginning (Edwards 2001). Local ARIGs first recruited new people and gained clients through web sites like Craigslist and online forums. ARIGs adapted and organized themselves to participate in the fast-changing Internet-dependent culture. The prevalence of free websites and index lists of paranormal groups increased the recognition of ARIGs in society, adding to their social acceptability, and made it far easier and discrete to report strange phenomena.

There was an investigation team right in your physical or virtual neighborhood. What used to be rather inaccessible is now a part of mainstream pop culture. It's never been easier to engage in paranormal research and investigation.

The relative privacy afforded by the Internet to research and consume related media for free allows ARIG participants to obtain considerable information while avoiding public exposure of their non-orthodox interests. Many ARIG websites routinely include instruction for how to undertake a paranormal investigation. Without any gatekeepers or editors to temper the approach, the ARIG attitude towards investigation is boundless. Characteristics of elusive entities are presented as factual and it's not uncommon to find information on protection against harmful spirits, demons, and aggressive beasts, though, as far as we know, no one has been killed by a ghost, demonic entities, Bigfoot, or an alien during an investigation. Or ever.

ARIG websites were a target of mockery from those with skeptical views due to their style and content. Many still are visually annoying or downright unreadable due to dark background, bright or flashing text (in red or green, tiny and gigantic), with moving image files, auto-play music or videos, and a general array of truly terrible web design. Even though professionally designed websites are more common now, many sites remain carelessly constructed with amateurish presentations of ideas and concepts, improper grammar, poor spelling, lack of punctuation and capitalization, and heavy use of idioms and slang. Sites regularly contain broken web links or sections that were empty or incomplete. The "results" sections of the sites frequently contained placeholders for further information that was not available but dated years before. With the exception of a few professionally designed sites, the majority of sites surveyed in 2010 were amateurish. ARIG websites nearly universally contained certain standard pages entitled "About," "Mission," "Evidence," "Equipment" and "Contact Us" and information about procedures and methodology, team members, and general paranormal subject information. Several sites had a space for introductory information about the subject area (e.g., "What Is Bigfoot" or "Ghosts 101"). Whereas scientific sites would commonly display abundant reference citations, this practice is rarely used by ARIG sites. Instead, multiple examples of copied and pasted information was on dozens of the same sites without attribution. Academic or scholarly science or historical references were virtually nonexistent.

ARIGs solicited new members on the Internet. Interest and enthusiasm, technical skills, and public relations are the main qualities solicited. Notably, scientific training or experience was not solicited. At the time of my study, no group seeking new members specifically requested science qualifications. Later, when I discovered a group asking for science experience, I answered the request. The ARIG leader told me I was the first person who applied with actual science credentials. Not many scholars or working scientists pay heed to paranormal topics, even in their own communities where ARIGs are active.

A group must have a spiffy name. ARIGs identify themselves almost across the spectrum with an acronym derived from their full name. What was often evident was that the name was derived from a creative, pronounceable acronym such as Paranormal Researchers of Odd Findings (PROOF) and Study of Paranormal Events Confirmed Thru Evidence, Research (SPECTER). Unfortunately, many groups appeared to put more time and effort into crafting their image, beginning with the name, than they did collecting, sharing and analyzing data and explaining results.

What goes up quickly, can come down just as quickly. The lack of investment required and ease of website creation also meant that group sites were abandoned with great frequency. The sites were put up and then not maintained or the domain registration was left to lapse after a year or two as the group fell apart or re-formed with new members (and a new name).

Since most groups are approached by those looking for help in understanding their own disturbing experiences, ARIG member often express sincere feelings towards assisting those individuals. Groups emphasize caring consideration for helping *clients* get through a difficult and confusing time in their lives. They frequently state that the clients' best interest is always the priority. The officially stated purposes, mission, or goals of ARIGs nearly universally expressed a commitment to understand the subject and to discover important evidence that leads to knowledge and acceptance of paranormal reality. Lofty goals are listed for their work: to be recognized by the scientific community, to provide indisputable evidence of the paranormal, to prove life after death, to help lost spirits cross over, to find Bigfoot, etc. An aspect of ARIGs that often goes un-cited by critics is their emphasis on community service—educating the public and raising awareness, fundraising for local cemeteries and historic sites, and advocacy for preservation of historic sites and nature conservation. This is both commendable and problematic as will be discussed later.

While a few groups notably listed one of their goals as "to have fun," most groups emphasized serious intentions. To exhibit their seriousness, ARIG members are visually portrayed on their websites in the process of conducting investigations, frequently using equipment. Humorous or comical photographs were rare. Group photographs commonly include members in matching clothing, frequently black, often with the group logo (earning them the derogatory "black shirt brigade" moniker from non-paranormalists). Cryptozoology or UFO field researchers may dress in modern outdoor gear fit for field work. Cryptozoologists typically require a hat and many sport beards that projects an "outdoorsman" attitude. Many ARIG sites showcase the group members. Short descriptive bios are often included which follow a typical pattern relating to the individual's day job or former career and a reason why each person is interested in the field. It was through the inclusion of their primary career information that I could ascertain that almost no participants in these groups are scientists by training or occupation (excluding computer science and engineering), but most often have "blue collar" jobs or no consistent or distinct area of employment. Another aspect that was clear in the members' personal info provided was the impact of personal paranormal experiences. Reasons for participation in an ARIG is commonly given as "curiosity," most frequently associated with personal experiences the individual has had that they consider paranormal or mysterious. They are searching for answers to their own questions. This is a strong motivator.

Across my survey of ARIGs, I observed a spectrum regarding a desire for publicity. There are certainly those groups that actively seek publicity or even promote themselves as potential television subjects. In fact, their desire to obtain their own TV show or to be featured on an existing one may be stated outright on their website. Publicity-seeking groups will list the press coverage they have garnered and showcase promotional video clips of investigations. Prominent ARIG representatives appear on radio shows, podcasts and in local interest stories. Some ARIGs will also promote their group via media contacts, seminars and events. Many also run web-based radio or video shows or podcasts. Some groups have their own YouTube channels to showcase video findings, photos, and sound files they say depicts evidence. This evidence is open to harsh comments by naysayers and other groups disputing the evidence or alleging fakery. Those groups looking for even more publicity and angling for their own TV shows will produce "sizzle reels" showing their most dramatic activity. Some groups will actively criticize other groups as publicity seekers, calling them buffoons or fakers or decrying their attempts to make the paranormal profitable. Several paranormal personalities obviously have attempted to grab fame and notoriety at every turn, whether that be by gaining a spot on a TV show or being the group that captures Bigfoot himself. This stated or unstated goal of fame causes friction and animosity as groups vie for credibility and attention from the public and media producers.

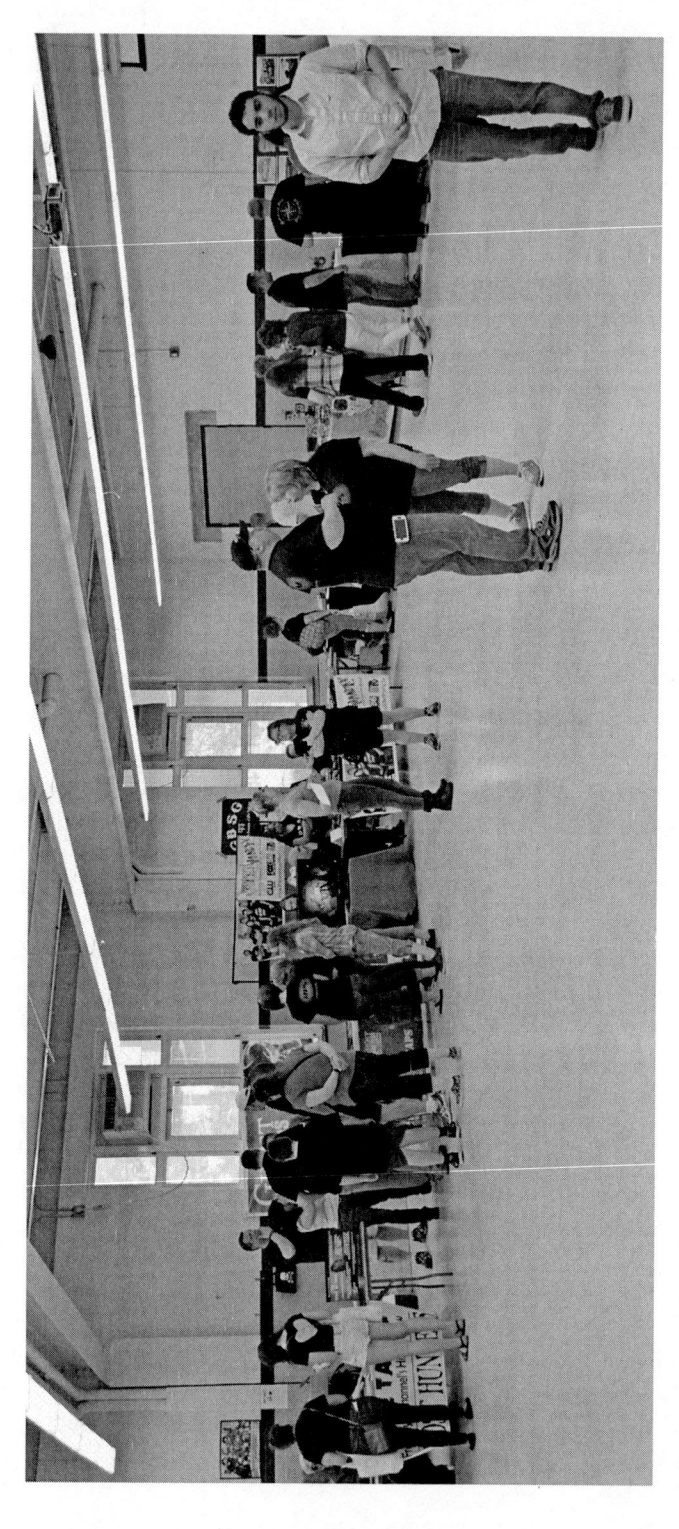

Most paranormal conferences feature vendors that sell books or ghost-related paraphernalia. The vender area also gives participants a chance to meet others and chat about the conference or items on display, and to share experiences. Taken at the Para-Unity Conference in Woodbridge, New Jersey, 2015. Photograph by Kenny Biddle.

Conferences

Paranormal investigators, cryptozoologists, ufologists and those generally interested in these topics meet up across the U.S. at conventions that host speakers, writers, television personalities, and vendors. Conventions range from small regional gatherings featuring local authors and experts to larger weekend gatherings that include international guests. Paranormal conventions ("paracons") are common and can attract hundreds of participants. Some paracons cross over all fields combining ghosts with psychic studies, cryptids, and UFOs while others limit the subject area to Bigfoot, for example.

The most numerous events cater to interest in ghosts and spirits. Vendors man tables to sell ghost hunting technology and paraphernalia. Book signings and paranormal celebrity meet-and-greet events, tours, movie screenings, and special ghost hunts are features of such events. UFO events are similar and national or international gatherings can attract major crowds. Some small local conferences combine Bigfoot with UFO topics into an event,[2] often to the chagrin of some vendors or speakers from one field who prefer no overlap with the other. Several large sci-fi or comic fan conventions include "tracks" within their huge programs to cater to specific sub-interests which include paranormal themes. Skeptic tracks may also appear alongside the paranormal tracks allowing the convention participant to sample a critical view of paranormal topics. Conversely, skeptical-minded attendees pepper the audiences of paranormal presentations, most often asking probing questions and calling out inconsistencies in facts and methodology.

Conferences and conventions are where ARIGs interact with each other, sharing techniques and findings and staying apprised of the latest information circulating in their individual communities. Conferences serve as critical face-to-face contact for ARIGs to form, add new members, and embrace new directions and ideas. Skeptical scholars are rarely included in the program. Hospitality towards skeptical views varies widely but, generally, conventions of ARIGs are geared towards those who subscribe to paranormal conclusions and are supporters or fans of the media personalities that espouse that view.

Professionalism

Sometimes, I think that ARIG participants may have taken the 1984 movie *Ghostbusters* too seriously and sound a bit too much like this "commercial" shown in the film:

DR. RAY STANTZ: Are you troubled by strange noises in the middle of the night?
DR. EGON SPENGLER: Do you experience feelings of dread in your basement or attic?
DR. PETER VENKMAN: Have you or your family ever seen a spook, spectre or ghost?
STANTZ: If the answer is "yes," then don't wait another minute. Pick up the phone and call the professionals...
STANTZ, SPENGLER, VENKMAN: Ghostbusters!
STANTZ: Our courteous and efficient staff is on call 24 hours a day to serve all your supernatural elimination needs.
STANTZ, SPENGLER, VENKMAN: We're ready to believe you.

The characters portrayed in *Ghostbusters* were PhD-holding professors in parapsychology. However, there have not been many professional parapsychologists, and the few university parapsychology labs that existed have been dwindling. Even fewer academic researchers have been able to shape their research into quests for UFOs and cryptids. A similarly small number of historians, psychologists, social scientists, and other scholars study the social aspects of paranormal belief. Notably, almost none of their work is positively recognized by ARIGs.

A real-life example of professional ghost busters does exist and shows how popular the concept of on-call parapsychologists was in the wake of the original *Ghostbusters* film. In 1989, the German Society of Psychical Research funded a project called *Parapscyhologische Beratungsstelle* (PCO) to have professionals and specialists available to members of the public that wished to report extraordinary experiences. They received 4,000 requests in a year for assistance, overwhelming the tiny staff and demonstrating an enormous demand for such services. Through 2003, the PCO received 2,500–3,000 requests by phone or in person per year (Von Lucadou & Wald 2014). I'm not sure if the volunteers were "ready to believe" but callers were ready to tell their stories to a person who would listen and not laugh or belittle.

Many ARIG participants and independent researchers write books. Between the rise of do-it-yourself publishing and the growth of niche publishers catering to the audience of paranormal reality television, publishing a paranormal book does not necessarily signify expertise nor promise financial gain. Regal refers to a handful of successful authors as "elite amateurs" who have the ear of the rest of the community as they provide opinions and hypotheses (Regal 2011: 17) yet their views remain outside mainstream academia.

A surprising aspect of ARIGs that I found was how often they invoke the term "professional" for their activities and deliberately eschew the term "amateur." Of the 1,000 ARIGs that I examined, 70 identified themselves explicitly as "professional" investigators with reference to the group or group members themselves. Reference to a "professional manner" related to conduct was not counted in the total as this was understood to be a behavior characteristic, not related to experience. What exactly do they mean by "professional"? A professional investigator is a tough job that requires a unique and practiced set of skills. Experience matters for those who work as investigators for law enforcement, insurance companies, or the government.

It could be said that certain investigators who have TV shows or write books might be considered "professionals" if they receive income from the activity despite lacking any formal training. Several para-celebrities charge fees for appearances and lectures. I found only one ARIG group that charged for their services, although the practice is becoming more prevalent but not advertised. Are they professional ghost hunters if they never actually find verifiable ghosts? How can you declare yourself to be finding Bigfoot when after many years, you don't have physical remains of one? Considering that the explosion of paranormal television shows was subsumed within the larger phenomena of reality television, unscripted works with primarily amateur casts or contestants, the use of "professional" to mean "gets paid for work" is colliding with the use to mean a "high standard of experience and/or quality." Adding "professional" to ARIG marketing material seemingly became a common means to express a sense of seriousness in their investigations.

The roots of quasi-professional paranormal investigation are in the foundation of the Society of Psychical Research (SPR). Work of the SPR was well known to Dan Aykroyd who co-created the movie *Ghostbusters*. According to Maher (2015), early SPR work relied on an observational technique. Their process was as follows:

- inspect sites to rule out natural causes;
- talk to the witnesses and make a permanent record of their experiences;
- obtain corroborating witnesses and supplementary information;
- attempt to observe the phenomenon or detect fraud;
- assess the data collected; and
- publish a formal report presenting the conclusion.

With this method, we have a reasonably objective attempt to document the activity and explain the potential causes. This method is not undertaken in full by ARIGs, most of whom are goal-oriented toward documenting what they will label paranormal activity by any means or measures. The alternative to an observational technique is an experimental technique which also developed out of the parapsychological community the SPR aided in creating. One example of this experimental technique outside of a laboratory was to provide a floor plan of a supposedly haunted location and, without providing any clues, have subjects mark out impressions. Again, ARIGs almost never invoke this process espoused by the original society.

Several ARIGs declare there are no real "experts" in paranormal activity because it's an evolving field. In other words, because no paranormal theory is accepted by most researchers, then no one can claim expertise. (See the chapter on *Science and the Paranormal* for further discussion.) With no set standards, and expertise and authority claims effectively neutralized, any researcher can call herself "professional." Expertise is democratized—anyone can be an expert if they can convince an outsider of their authority. A potential client may see the word "professional" and assume there are some special qualifications for being a ghost hunter or Bigfoot investigator. There isn't.

Ghostbusters was just a movie. It wasn't real. Not only do some people think their local ARIG can pop over to deploy a ghost trap and remove the problem (Krulos 2015), some ARIG members aspire to be such experts.

Educating the Public

ARIGs commonly state a mission of education, which means to provide information they hold to be true or find useful regarding their particular view of their subject matter. The IIG survey (Duffy 2012) reported that 85% of responding groups of paranormal investigators consider "educating others" as a main goal after "helping people" and "conducting research." About a third teach classes on the subject and one in six reported they give tours to the public. Yet, is what they are teaching factual or is it promotion of a belief? Are they qualified to teach about ghost hunting or cryptozoology, subjects that, in a broad sense, are multi-disciplinary and include topics typically covered in academic settings such as history, folklore, and science?

Appeal of participating in an ARIG includes the status that comes with being part of an organized group and having a personal title like "President" or "Chief Investigator." In society, honorific or official titles like "Dr." or "Professor" often require academic or professional credentials. Titles elevate the spokesperson's apparent importance and suggest experience or professionalism. This is effective in media. Researchers adopt a confident tone of a topic expert to the press, emphasizing their dedication to their field as a demonstration of reliability and seriousness.

Several ARIGs arrange one-day or weekend training events or workshops to teach methods of investigation to anyone who is interested. These include "ghost hunting basics" or camping excursions to search for Bigfoot evidence. There is often a charge for these programs. Many ARIG representatives (and individual paranormal researchers) will seek more formal and informal educational opportunities such as lectures, presentations to the public, guided tours, and even formal classes online or in person at a school or community center. The same tone of expertise is invoked in those situations. Across the country, ARIG leaders partnered with local community centers and small colleges to offer continuing education classes (without academic credits) in paranormal topics. ARIGs and other community individuals are invited or allowed to present because the institutions providing community education look for subjects that are popular and bring in people to the classes, some of whom pay or may pay for other classes in the future. ARIG members can be found leading classes in paranormal investigation, developing your psychic powers, or guiding you on how to talk to your guardian angel. Opportunities to lecture and teach are highly sought by the ARIGs as perfect opportunities for educational outreach and promotion of their conclusions, beliefs and methods. The appearances might result in recruiting new members, donations, or increased access to prime investigation locations. Any semblance of affiliation with a school, college or university is readily used in the group's promotional materials. While use of and association with the school appears to be a tacit endorsement of legitimacy for the ARIG activity, it's almost never an official endorsement by the institution. The few college courses for credit taught by professors that explore fringe subjects are often explorations of folklore, mythology, and cultural history, or exercises in critical thinking and identification of pseudoscience. There is value in studying, for example, the history of ghosts (see Finucane 1996) as a reflection of cultural norms. The exploration of the growth of UFOs' popularity in the U.S. nurtured by an American sense of distrust of government and science is of value from a social science and cultural perspective. A few formal academic programs for parapsychology remain, focusing not on how to use gadgets or how to exorcise spirits and entities, but on how to design a useful psychological test and apply proper statistical methods in analyzing experimental data—entirely different from what is presented by amateur researchers. These courses generally do not involve attempting to interact with spirits. Are ARIG-led classes and educational events legitimate and useful? They certainly *could* be. For example, a class could be structured with reliable and referenced information to provide a worthwhile educational experience. A lecture can inform an audience about the history or current popularity of the subject. Providing a class or presentation equivalent to a formal or informal educational opportunity would take a considerable effort and teaching skills that not many hobbyists have. True experts in any field are expected to have several years' worth of comprehensive knowledge of the topic. That knowledge isn't used to impress or gain attention. It is digested and used to feed further understanding. Public ARIG presentations I've observed lacked authenticity and deep knowledge. Key words were used without a coherent meaning provided. Speculative ideas were thrown about without references or foundation. Expertise involves more than regurgitating names and facts; it requires comprehension and application of that knowledge that results in a capability to explain. Expertise allows for sound judgments to be made and the ability to think critically about the primary issues and problems in one's field. Gaining that level of expertise is possible by amateurs but it takes decades of dedication. To take the position of an authority and educator is ethically questionable when expertise is shallow and contrived.

How-to ...

In addition to public outreach education, ARIGs compose their own handbooks and how-to guides, often publishing them as books. For these fields of study whose advocates openly state that there is not one "right" way to do things, there exist a great number of books on how to do such investigations. With some exceptions, these books are written in an authoritative voice but are self-published, lacking an editorial hand, of poor quality, and replete with scientific inaccuracies, outright mistakes and outrageous assumptions. It's best to provide some examples. In one of the more readable, if logically vexing, handbooks Joshua Warren perplexingly states: "Ghostly activity is the most efficient link to scientific study of the concept of an afterlife" (Warren 2003: 115–16). This is news to the scientific community. Though the statement makes no logical sense, it does reveal the author's belief in life after death. This belief tints the entire book which is marketed as an objective "how to" guide. Gibson, Burns and Shrader (2009: 43) discuss EVPs (electronic voice phenomena) as the most compelling evidence one can gather on an investigation since it shows "solid concrete interaction" with the paranormal. There is no scientific basis for such a claim and it conflicts with academic and parapsychologists who do not consider EVPs to be worthwhile evidence of anything. Formal research into EVPs is sparse and not promising (Leary & Butler 2015). To consider it as a form of ghost communication is a belief, not a fact. This same source (which is aimed at a teenaged audience) also suggests that psychics deserve notice since they are "right sometimes" (Gibson et al. 2009: 49). Such suggestions to young people forming their worldview are misguided. We all guess right sometimes; to jump to the conclusion that psychics should be taken as credible tools to interpret a situation via undiscovered sensory means goes far beyond what we know about nature. The authors, who claim a mantle of authority but have no academic credentials, give potentially science-minded readers an erroneous idea of the process of effective investigation and inquiry. A detailed review of these and other ghost-hunting guidebooks is presented in the Appendix.

Many ufology and cryptozoology-themed books are, likewise, marketed to young readers and those without any scientific background. Typically, these books repeat unverified stories of famous "classic" sightings and eyewitness reports contained in previous books. Most mass market paranormal-themed non-fiction content provides mysterious stories without references, context, or critique. This is more akin to folklore than fact. Authors propose speculative and colorful "theories." Though readers find such ideas interesting and entertaining, this is not good scholarship and these volumes are unreliable information. Guidebooks specific to cryptozoology include speculative details about potential flesh and blood creatures[3] hiding from humans but leaving traces behind. A slew of self-published books exist that describe each author's experience with cryptids and suggests the best ways to bring them close. Field work in cryptozoology typically means being able to spot anomalies in nature such as disturbed vegetation, constructed or manipulated objects (like branches or rocks), tracks and spoor. Regardless of how many "field guides" to find Bigfoot, lake monsters, and other cryptids are produced, no mystery monster will be discovered in this way. Such field guides oversimplify biology and ecology. The cover summaries promise that the creature you are looking for is out there to find if you just use this book. Ecologists and zoologists, trained to detect and document the presence of animals and to understand ecological systems, aren't looking for cryptids. New species are more likely to be found among disregarded museum samples than out in the American forests. As with all other topics, you can find an array of opinion blogs, websites, forums, Internet radio and podcasts available on cryptids and supposed encounters ranging from outrageous fiction to science-based discussion by experts. Referenced, expert, zoology-based content for Bigfoot and other cryptid seekers is uncommon.

UFO-related handbooks for investigators are rare, however those that claim to be research guides are almost universally written by authors who hold belief in the extraterrestrial hypothesis or that the government is covering up information. Podcasts and Internet forums such as *Above Top Secret* keep researchers and enthusiasts up to date on news and rumors.

Reference books on fringe topics are available in a range from juvenile literature to dense, dry scholarly works that contain impenetrable prose and in-text citations. Some popular guidebooks suggest the best places to look for UFOs. Atlases and compendiums of haunted or "weird" places have sold reasonably well, likely boosted by television exposure to such places. An excessive number of paranormal nonfiction books written by amateurs for amateurs flood the market. The more scholarly publications produced by academic presses cost more and are held usually only by university libraries, limiting their usefulness to non-academics. ARIGs may sometimes derive their techniques from literature sources but it is more evident that their processes are based on following what they have seen on TV shows.

Social Aspects and Potential Harm

The social aspects of being in an ARIG are substantial, as I alluded to several times. Being part of a specialized group feels unique, exclusive and sometimes provides an extraordinary experience (Childs & Murray 2010). The participants can, without obtaining a degree of higher learning, become self-appointed authorities in a chosen field (Northcote 2007). Participation in paranormal themes and activities can be personally rewarding and is considered a "lifestyle choice" for many ARIG participants (A. Hill 2010).

Socially marginal people are less likely to gain rewards from participation in organizations to the same degree as powerful members of society. They aren't as likely to become leaders or be influential. But within an ARIG that anyone can start on their own, leaders bestow responsibility and titles to themselves—a ufologist, cryptozoologist, paranormal researcher, or a ghost hunter, as if it is equivalent to a profession. They have a sense of camaraderie with their fellow members as well as feeling at ease with others that share a unique interest without fear of ridicule. As people invest time and effort into a subject they are passionate about, they feel an increasing emotional investment (Stebbins 2007), perhaps even a sense of responsibility to pursue the topic. Fellow members can reinforce or enhance prior belief—Bigfoot is out in the woods and needs to be protected or alien spacecraft is visiting earth. Once a person becomes emotionally (or financially) invested in a field, it becomes increasingly difficult to extricate themselves from that circle even if evidence accumulates that you are on the wrong track. The "group" aspect can have an additional cohesive effect for keeping people involved in the subject, or it can lead to toxic dynamics within the group as members become disillusioned with the leaders or each other and exit.

The ARIG participant may be seeking not only paranormal explanations but larger meaning in life (Bader et al. 2010; Northcote 2007; Booker 2009; Dolby 1979). Because of mundane jobs or lack of opportunity in life, we may find ourselves feeling unimportant or uninspired as society fails to provide a basic psychological need for a sense of importance and individuality. Sociologists recognized that an anonymous life, mundane job or missed opportunities contribute to *ennui*, a directionless feeling of unimportance. Wishfulness that a long-standing mystery will be solved or a desire for an experience is a strong impetus for ARIG participation. Acquiring what feels like hidden knowledge is empowering and, for some, is incorporated into how they construct meaning and make sense of the world (Jenzen & Munt 2013: 34). Religion scholars have recognized that such context can draw people towards alternative beliefs (Bader et al. 2010: 55). The same draw may apply to ARIG participants who seek a unique and powerful experience (Clarke 2012, Hess 1993; Bader et al. 2010).

Ghost hunting as leisure activity existed as early as 1858 (Clarke 2012) though ghost stories have been part of our culture for thousands of years. Amateur research into paranormal topics is generally harmless to participants and their families (Potts 2004) but there are exceptions. Most ARIGs are diligent about obtaining permission to access property but a few careless curiosity seekers have engaged in unlawful trespassing and defacing or damaging property.[4] Unfamiliar locations, restricted and remote areas, and abandoned buildings are rife with hazards, especially at night. In several instances, trespassers have been arrested, injured, and even died trying to experience the paranormal.

ARIG sites mention, almost apologetically, the substantial costs (in time and money) associated with travel, undertaking investigations, and purchasing and maintaining the necessary equipment. ARIGs typically do not charge for personal investigations but may accept donations or do fundraisers to cover costs. Duffy (2012) found a median value of $1,000 spent by paranormal investigation groups per year.

While there are a few unscrupulous characters in the Bigfoot, ufology, and ghost hunting communities out for personal gain, most ARIGs will expel those that cause trouble for others in the group or are participating for the wrong reasons, faking evidence, or otherwise threatening the integrity of the group.

Indirect harm is possible when individuals become so involved that it interferes with a healthy lifestyle—financially, psychologically, socially, or physically. This is not unique to paranormal pursuits. A common pitfall is spending too much time on the activity or exorbitant amounts of money on supplies, travel, and equipment. Time and money are invested gaining worthless credentials from training classes or attending events. As with any hobby, there a risk that the participant goes too far, neglecting real world issues and making the activity an escape from reality, conformity, and routine. If they believe a breakthrough is just ahead, their ego may prevent them from letting go. Investment in belief to the degree that its importance overwhelms concern for health, safety, and providing for family is a possibility. Relationships and financial assets have suffered by misplaced attention to ARIG-related activities or individual determination to find "proof." Krulos (2015) illustrates the potential relationship problems that may result from one partner's involvement in a paranormal hobby to the dismay of the other. A textbook example of such a person was one of the most colorful and memorable Bigfoot hunters ever, Rene Dahinden (Regal 2011). After 40 years of searching, he had little to show for his efforts. "You know, I spent over 40 years—and I didn't find it," Dahinden said to Christopher Murphy. "I guess that's got to say something."[5]

The belief in paranormal activity may be generally harmless but can also lead to association with dangerous beliefs such as curses, possession, or conspiracies which can be life-altering. A paranormal enthusiast can become so enveloped by the subject they can believe an evil spirit has followed them home, that Bigfoot is communicating especially to them, or that they are being watched by government operatives. Fringe beliefs may influence how individuals think about federal policy and laws, and how the government allocates funds for environmental protection, national security, or space exploration. Paranormal concepts become part of a worldview which can influence how the participant thinks about and behaves in other aspects of life.

The most serious threat of all are those on the receiving end of ARIG involvement who have mental health or domestic issues. Most of the ghost investigators I've spoken with have stories of clients that are clearly in need of mental health assistance and do not have a solid grasp of reality. Those who state they believe they are being haunted, visited by Bigfoot, or abducted by aliens may have underlying problems that amateurs are not equipped to handle. ARIGs typically do not have access to professional psychologists to which to refer clients. While many ARIGs will steer clear of these types of situations, data is not available regarding how often investigators will exacerbate an illness or situation or cause the client to delay getting proper help.[6] There are some ARIGs who attempt to screen claimants to determine if they are hoaxing or in need of other kinds of help (Krulos

2015). Any attempt by ARIG members to provide counseling or therapy is unethical and potentially illegal under certain laws against unauthorized medical practice. Especially in a situation where children are involved, ARIGs tread dangerously in advising clients and can be liable to legal recourse should something go awry. Considerable harm is created and situations worsened by ARIGs who tell distraught clients that they are plagued by demons, spirits, Bigfoot, or alien visitors. What if ARIGs miss an abusive scenario by misinterpreting it as a haunting or a possession? Anyone who diagnoses demonic possession and attempts exorcism risks mental and physical harm to the person. For ARIGs to involve themselves in these kinds of dangerous and emotionally-charged situations thinking they can fix what's wrong is the height of hubris.

7

Science and the Public

The gap of understanding between what amateur groups think it means to do science and the standards and goals that exist in the professional scientific community is unsurprising considering the paradox that Americans have with science. We admire it, we know it's useful, but our grasp of how it works is weak. People like science ... when it's on their side. Based on decades of polling and surveys, especially that of the ongoing Science and Engineering Indicators reports produced by the National Science Foundation, as well as public feedback and cultural products, we know the general public does not understand scientific purposes, values, and processes. Most Americans have a difficult time explaining what good science is at all and accept a great deal of false or bad science as reliable. Instruction on formal science understanding and, more importantly, appreciation, is lacking in today's typical school curricula. Most people don't have a working scientist in the family. When faced with decisions in life that are (or should be) informed by science, the average person may not feel comfortable with accessing and deciphering scientific data and conclusions since most of the science we get comes from the press and media (Nelkin 1987). As a side effect of our lack of familiarity with good science, the public can be easily fooled by imitations of science. It's a common marketing ploy to feature sciencey-looking things to sell products and services. A sense of science "authority" can be easily conjured by just a person in a lab coat (Toumey 1996).

The gap between scientists and the public opened in the earth 20th century when "scientist" became a specialized profession and required extended study and a different lifestyle. The scientific community talks mostly to itself and has remained insulated from most public exchange. Unfortunately, that created a detrimental disconnect between science and the public. Scientists don't talk to the public about their day-to-day work and its consequences very often or very well. C.P. Snow[1] issued a sentiment in 1959 that still reverberates today, stating there were "two cultures" in academics—sciences and humanities—that couldn't communicate with each other. The public sees "two cultures" across an even deeper chasm—academics versus non-academics or scientists versus amateurs. In the late 20th century, the facts produced by science and research became detached from the process that produced that knowledge. In popular media, especially television, we do not see represented the stringent method and specialized processes required for scientific conclusions (Toumey 1996).

Science progresses on a path quite different from what the public sees. Regular surveys about the public understanding of science tell us that the non-scientist doesn't comprehend well the importance of critical concepts like controlled trials, peer review, skeptical criticism, and holding provisional conclusions. People form their ideas about science from the input they get via basic education and popular culture. The scientific news and literature is very different from news and media meant for public consumption and it is not easily accessible to the public. Most of us get our news through sources that assume a non-specialist audience. These emphasize exciting or monetary aspects of the latest findings in science, disregarding or oversimplifying the amount of work necessary to reach the conclusion. This framing of science news, unfortunately, misrepresents the process and skews the public understanding of science.

Public ideas about science have changed drastically over time. In the book *Science Talk* (2007), Daniel Thurs describes how Americans call upon a reservoir of words, images, and ideas to define the nature of science. He notes that it is important to distinguish between how the public defines science and how science defines itself, which are very different things. In public discussion, science talk centers on the authority, boundaries and outcome of the process, not on the philosophy and methodology of science itself, the latter being concepts not generally emphasized in general schooling. Most Americans give little thought to science philosophy because they don't need to. They won't practice science but are interested in the results and benefits to them. The public idea of science is constructed by the public itself to suit its needs. So we would expect to see the popular cultural image of science/scientists as inconsistent with the scientists' version.

In scientific research, definitions are of utmost importance to communicate clearly and understand precisely what is being said. Incongruity exists between casual usage of words and academic contexts. Some obvious examples include the common use of the terms "theory" and "energy" which have very specific meanings in scientific fields and very different connotations in general English usage. Before discussing how amateurs use science in their investigations and research, we must first establish the meaning of three basics concepts—science, the public and the scientific method.

What Is Science?

"Science" is a bit hard to define. The meaning, like with many other words and concepts in language, changes through time and is dependent upon personal values. If you are one who values science as a reliable way to understand the world, you likely have a different and more stringent definition of the term than someone who values it less. People project their own meanings onto science based on their life experiences. Non-scientists typically see science as monolithic, consisting of facts written in books. Science (with a capital S) represents orthodoxy, it is unchanging and authoritative. Science is perceived as being complicated and specialized, intrinsically difficult for a layperson to grasp (Toumey 1996). Admittedly, it is complicated. In today's modern western culture, *science* is often connected to or interchanged with *technology* (Thurs 2007) as the American public connects the process of science with the useful outcomes, products, and processes that make our life better. Technology is a vehicle for dispersing science in our society but we view in in light of products we deem relevant.

"Science" is both a process that generates knowledge and a collection of systemized knowledge (Ziman 2000). Early use of the word "science" referred to this collection aspect rather than the process. In the 19th century, "science" was restricted to referencing academic work, mainly regarding nature (Pigliucci & Boudry 2013). But in modern culture, we talk about the "science of" something or about "doing science." Therefore, when referring to science, we might mean the derived body of knowledge, or the specific approach taken to obtain that knowledge. Science is a process, a way of looking at a topic, a community, an infrastructure, a career, a set of results, an authority, a media category—we can use the word in so many various ways, which means it can be abused in just as many ways.

Does science belong to everyone or is it the domain of specialists, the scientists? This is a complicated question that does not have a simple answer. *Real Science* by Ziman (2000) is a readable reference that breaks down science into its components and describes why they are integral to each other. Ziman describes how almost all science is based on observation. Experiments are acts of observation designed to produce specific data that inform a question. Use of instruments in a situation can be an experiment. Therefore, the ghost/ARIGs using EMF meters or other devices are conducting an experiment. Rapping on a tree with a stick to test for a response (from Bigfoot) can also be considered an experiment. To be useful, the experiment must have context and a clear purpose. Experiments are desirable but not always required. Collecting observations as statements and facts such as in historical investigations is also part of the scientific process.

Humans collect information by personal experience. Or, we can gain information indirectly by watching or reading about a subject. Evidence we gain directly via our senses is called *empirical* evidence, in contrast to evidence we gain from reasoning or intuition. Science is dominated by empirical evidence. A critical caveat to empirical evidence is the question of the credibility of that evidence. Was it accurate? Was it perceived and recorded correctly? Can it be reproduced reliably by others? Will the same results occur as expected the next time? Observations by witnesses, even trained witnesses like scientists, can be biased by various factors. Therefore, citing empirical evidence may not be as clear and direct as we assume (Dewitt 2004). Potentially confounding factors must be made explicit as part of the reporting on the experiment or observation to fully account for them. Examples of basic confounding factors include: it was dark, there were many people around, the weather was poor, there was no scale for size and distance measurements.

Inherent to an observation being scientific is that it is objective. Quantification of observations is one way of making the data objective. We measure precisely, then the next person who measures should come up with the same results, within a reasonable margin of error. Separate observations that result in the similar measurements strengthen the evidence. Observations are also heavily influenced by the observer's preconceptions (such as a commitment to a certain explanation) and the behavior of people around the observer. When you are told to expect some force to be at play (an object to move), you are more likely to perceive it moving. If surprised and frightened (such as in a car crash or earthquake), reaction and observations can be skewed toward personal preservation. Scientists are trained to be better observers, to focus narrowly on what needs to be recorded and how exactly to record it to minimize errors. Since we are human, errors can't be eliminated entirely. Error minimization is addressed via the scientific tenet of communalism (discussed ahead). From the setup, through findings, to the conclusions, the process must be transparent and available to others to attempt to reproduce. The biases of any individual investigator or researcher that affect objectivity of the observation is "neutralized in the collective outcome" of scientific debate (Ziman 2000: 159).

In *Philosophy of Pseudoscience* (Pigliucci & Boudry 2013) the contributors stress that science is a "collective enterprise"—its strength and privilege derived not from the individual theories or models but in the established properties of the community and the interaction of the members of that community. The scientific community is responsible for sorting the wheat from chaff, building upon others' sound work, or to provide informed criticism expecting to be answered. Shortly after its formalization, science progressed to a level where one individual can't do much, but the community can collectively solve vexing questions by cooperating and providing feedback (often corrective or critical) to gain progress. Coordinated efforts, collective sharing of research results, and reinforcing findings are essential to establishing reliable knowledge. Ziman (2000: 110) states emphatically that research results do not count as scientific unless they are "reported, disseminated, shared and eventually transformed into communal property by being formally published."

As children, we likely obtained our first ideas about science from pop culture. Most of our exposure is based on stereotypes of scientists we saw on TV or in movies—un-emotional, serious, socially separate, intelligent men in white lab coats. Scientists spoke a complicated language, comfortably handling equipment and gadgets, taking notes and recording data. Scientists were often seen as having hidden knowledge in their labs and notebooks that only they understood. Application of scientific methods isn't a typical way of thinking about life. It's not common sense. Consider the observation we make daily—sunrise and sunset, which is most often depicted as the sun moving around the earth. This is incorrect but without being taught what really happens, we might not know or care about the distinction. Sometimes, the empirical truth wouldn't matter in our day-to-day lives. We don't live scientifically, testing our hypotheses in a controlled and formal way and then asking others to weigh in on it and point out our flaws. Science requires a degree of rigor we are not used to applying, but we do use a loose approximation of a scientific or rational method for some decision-making processes to determine what is the most likely true answer, especially when we know there already is a body of reliable of information we can check.

The key to science's air of exclusivity and power in modern society is related to how well the borders of science can be defined and held tight. Borderlines that have been set up around the idea of science and scientists are revealed in the use of terms such as "the scientific method," references to the "ivory tower" of science, and the distinction between genuine science and pseudoscience (Loxton & Prothero 2013) which will be discussed in a Chapter 12.

What Is "the Public"?

"Science" was tough to define but the "public" is even more nebulous as there are multiple "publics" depending on the subject. Generally, we can say that "the public" consists of informed citizens who are participating in society (Gregory & Miller 1993). Some groups choose to pay attention to different parts of society or to ignore them, and different sections of the public come prepared differently to each subject area. Some members of the public have lay expertise that may be equal to but separate from professional expertise. For example, consider the farmer and the agricultural scientist: one knows practical applications and the other knows detailed formulas or conceptual models. Regarding ARIGs, consider the astronomer and the ufologist, the zoologist and the cryptozoologist, and the psychologist or physicist and the paranormal investigator. ARIGs may appear to have expertise in comparison with the general public but likely not have comparable expertise to scientists and professional experts who may also consider themselves part of the public sphere for discussion and decision-making. When it comes to areas like children's education, community decisions, or national policy, various parties (academics, social scientists, religious officials, parents, business people) assert their opinions on the basis of their sphere of experience. Such pluralism is almost always a positive contribution for society.

Another view of the "public" is that of the "mass culture." "Mass" suggests vast numbers, undifferentiated but heterogenous in make-up. "Mass media" is consumed by the majority instead of a specific group.

Science educators have often taken an approach to the public in terms of what is called the "top-down" or "deficit" model. The "public"—defined in the broadest, shallowest terms possible—consists of individuals who are empty vessels that need to be filled with scientific facts and ideas. Then, the assumption is that the public will use that information to make scientifically informed decisions and society will be more science-literate. Considering the diversity of the public in background knowledge, interests and values, this model is not realistic. The simplistic model of delivering facts doesn't work. The approach to education must be contextual. Information must be delivered in a way that is useful and meaningful to the audience or it will likely be discarded or ignored. The many varieties of the "public" hear what it wants and needs to hear.

The Scientific Method—There Isn't One

In school, we are taught the *scientific method* as a step-by-step action of hypothesis, testing by experiments, then theory formation. We are delivered an idealized process of how science works (Dolby 1975). This overly simplified procedure is not an accurate reflection of how science progresses in the real world. There are reasons behind the many stringent rules of scientific research. The execution of the process, in accordance with those rules, is often tricky, with nuances and complexities.

Once upon a time, "scientific method" was not part of the common vernacular. When the term began to be used in the mid-19th century, it was synonymous with "thorough" and "careful." For many of us, our first introduction to science may have been in elementary school, when we were taught the "scientific method" to investigate nature that went something like this: observe and gather facts; derive the question you need answered about those facts; propose an explanation for the facts that answers that question; test that explanation. The scientific process is equated to the hypothetico-deductive method. Observations result in formation of a *hypothesis*, a prediction is deduced based on the hypothesis (or multiple working hypotheses), and these hypotheses are checked for accuracy, and then supported or rejected. Only natural explanations are assumed (Pigliucci & Boudry 2013). In an idealized lab setting to test a chemical or physical question in a controlled environment, this recipe for investigation would work. But when you move out of the lab, into the very messy and busy real world, it is impossible to control all the variables that can mess with the results. Think about all the possible factors at play in a situation where a witness experiences something strange outdoors. A smell or noise could be caused by countless things, some of which we don't readily suspect. We misinterpret information relayed to us via our senses. Then there are the many and various environmental considerations that we are not measuring or do not know are operating. Investigating an open system is incredibly complicated and difficult (Pigliucci 2010). Certain types of investigations do not lend themselves to experiments but to observation and pattern-seeking. Some research is like detective work, putting the factual pieces together to form a reasonable conclusion. Good science can be done this way without a formulaic and idealized method (Pigliucci 2010).

There is no foolproof, formulaic recipe for inquiry that can be applied to all subject areas because the circumstances of the phenomena to be studied vary. Even when diligently using a prescriptive process in a laboratory setting with controlled variables, the results can be wrong. The scientific method is more of a mindset than a formal technique (Collins 1987; Pigliucci 2010). The formal training a scientist receives and practices in conducting research builds this science-oriented mindset.

The idealized scientific method assumes one person or a small group are working to solve a problem or answer a question and that there will be some "Eureka!" moment. That scenario is rare in today's world of complicated problems where major research is conducted by large teams. Another larger set of scientists offer their questions, opinions, suggestions, and complaints about the research results. They will pick every nit and they will expect the results to be replicated. An oversimplified cookbook method ignores this critical social aspect of science (Lyons 2009) that makes it the most reliable means of knowledge-production humans have. A scientific effort takes time to accumulate close to true or probably true models of nature. These models, called "theories," are strengthened through additional empirical evidence, observations, and by testing to determine if additional conclusions of research fit the model or not (Haack 2003). The idea of a single iconic "scientific method" can be scrapped. Yet, for purposes of this volume, the scientific method still functions as a trope, a useful rhetorical device that, in this case, represents being careful, systematic, and accurate.

Why Science Is Privileged

As stated, science is a process, a body of knowledge, but also a social institution—a complex system where the people doing research, their instruments, institutions, and journals all interact to produce knowledge. Science is not one action or experiment; it is made up of coordinated actions in support of a larger scheme (Ziman 2000). There are other ways of gaining understanding about the world but, in modern society, science is a privileged method of inquiry. Why is science inarguably the most reliable form of gaining knowledge? Let's consider the alternatives. You can accept everything people tell you. That usually works out fine in everyday life but when dealing with complexities of nature such as figuring out physical interactions, if a medical treatment really works, or what exists out there in the universe beyond earth, testimony from your friends and family fails. Some people say they have personal revelations and that you should believe them. They maintain that these judgments are as valuable as scientifically gained conclusions. Revelations are interesting, but people can and do make up nonsense and say "they know." Or they make mistakes and adopt a baseless philosophy. So personal epiphanies may be wonderful and gut feelings and intuition useful for the individual, but such ways of personal knowing are not reliable for the rest of us. Science is the most reliable because it addresses and attempts to eliminate all the ways errors can happen. There are countless ways that we can be fooled by our observations, calculations, and conclusions. Scientific claims are testable and scientific theories are reliable in that we can use them to predict what will happen next or where to go looking for the next discovery. The ability to *predict* is the strong backbone of science. If a theory can't help others explain and predict how nature has worked in the past and will likely work in the future, we can't gain useful knowledge and progress in our understanding.

ARIGs commonly exhibit an unusual love-hate relationship with science. They love its social prestige but resent that conventional science as a body rejects their non-materialistic ideas about magic and mystery in the world. They often express anger that they are excluded. In a bit of "sour grapes," they berate science and even sow doubt about science in general and its role in society. Amateur investigators state they feel they are more true to the scientific endeavor of seeking the truth, and are proud of their lack of credentials, because an academic path is stifling (Regal 2011). This view is a way of rejecting official scientific expertise, but not the process of science itself (Blancke et al. 2016). They appear as underdogs, a socially useful position. Resentment against orthodox science is palpable in the words and deeds of paranormal seekers like Ivan Sanderson, Harry Price, and Rene Dahinden. A commonly-used retort against science by those who espouse a paranormal or unconventional conclusion is "science doesn't know everything." Of course, "it" doesn't. Science is done by people, and people are limited in time and effort. That doesn't mean science is a bad thing. Consider that we have a very safe and productive modern society based on scientific advancement. Those who degrade the value of science may be expressing frustration that they don't have solid, accepted, credible backing for their own unorthodox ideas. Science can be scapegoated as the bad guy because scientific laws and conclusions undermine cherished belief and beloved ideas. The logical, objective scientist is framed as the buzzkill. Science as a process isn't perfect but it's the best method we have. Working within the purposeful constraints to develop valuable results provides rewards. A free-for-all method where anything goes will not work to progress human understanding of nature.

Science Becomes Exclusive

In human evolutionary history, people were too busy surviving to develop more than cursory ideas about how the natural world functions. The development of science required leisure time. In the days before science was an actual profession and "scientist" was a real title, fantastic stories about the world were taken at face value (Regal 2011). Naturalists appeared around the 16th century. Passionate about their subjects of study, naturalists most often held the view that the world and everything in it was created by God.

Modern science has its roots in amateur activities prior to the 19th century (Mims 1999; O'Connor & Meadows 1976; Ziman 2000). Popular ideas about science evolved significantly since the word "science" came to be. At first, it just meant a body of reliable and systematized knowledge. That general way of referring to "a science" was in use until the early 1800s.

When "scientist" became an actual profession—that is, when the complexity of the process and knowledge grew to the degree that specialized training was required—amateurs were pushed out and a unique language and specific structure of science developed. By the 1870s, scientific discussion in the U.S. had passed beyond public understanding, but the utilitarian need for scientific research was clear (Daniels 1971). In the early 1900s, scientific fields were distinguished as professional classes of participants—geologists, botanists, agronomists, chemists, physicists, etc. Science became a distinct activity unto its own. Language, credentials and institutions actively built virtual boundaries around science resulting in a sanctioned procedure and formal set of knowledge. Science as a discipline and community was established far out of reach of public understanding, yet scientists found this disconnect to be of no concern as they cared more about the approval from their professional colleagues. Peer acceptance was all that mattered.

Constructed boundaries enhanced the reputation of science as a distinctive (perhaps "honored") way of knowing about the world and excluded that which wasn't science (conveniently judged by the scientists themselves). Boundaries were constructed by institutions of learning but also by science journals. Research was not science unless it was published via this specific route (Thurs 2007). William Whewell, who coined the word "scientist," opined that amateur contributions were not valuable. Facts must be attached to theory and theory was the realm of the professionals (Lyons 2009). Academics pushed amateurs out of obtaining funding to do research. In the 19th century, government scientists picked up the research into their country's plants, animals, rock, fossils, and historical artifacts. Those who were not associated with institutions were left out (Regal 2011). As the scientific community organized into an "establishment," an ethos developed. Certain standards of practice were expected of a "scientist," foremost of which was the gateway that existed exclusively through higher education.

Scientific discoveries contributed to human societies in (mostly) positive ways; therefore, the prestige of being "scientific" grew. "Scientific" was associated with being more true and reliable. The biggest drawback of this prestige derived from the rigor and professionalism of science was that the scientific community itself and the capacity to understand how science really worked receded ever farther from the grasp of the non-science public (Denzler 2003). Being a scientist was special because not everyone could do it. Being scientific was a high standard. Science is not easy to do. Today, being affiliated with a scientific institution (or even what sounds like one) is a powerful public persuasion to being accepted as legitimate. "Scientist" has become a social class or caste isolated from the rest (Toumey 1996). Disciplinary labels like "physicist" or "psychologist," and titles and letters after names are rhetorical flourishes of credibility indicative of expertise. Association with a scientific institution discourages questioning from outsiders. Framing of a speaker with impressive sounding affiliations works to increase their credibility, warranted or not.

According to Daniels' *Science in American Society* (1971), the American public adopted a "childlike faith" in science. Because of the lack of familiarity with the process of science, society can blindly accept any claim too quickly if coated with the patina of science. Charlatans and incompetent scientists exploit this public credence in *sciencey* approaches.

On the Fringes

More than a century of organized science has resulted in a narrowly focused research agenda for many scientists. Scientists have become extremely specialized by necessity to dig deeply into a complex and narrow subject area. Expertise in one area of science in no way guarantees competence in even closely related areas. For example, double-Nobel prize-winning chemist Linus Pauling promoted medical claims regarding ingesting massive doses of Vitamin C. Pauling was a highly esteemed scientist in molecular chemistry, but not in medicine. There is a term for those who win Nobel prizes but, perhaps due to ego, go on to stumble badly in other fields: "nobelitis" (Diamandis 2013). Knowledge in one specialized niche can fail to translate as well to another niche, especially a complex subject with as wide a scope as paranormal subject areas. Fringe topics are of a different flavor than orthodox science topics. For example, cryptozoology is not simply a branch of zoology. It is an inter-disciplinary field formed by threads of folklore, psychology, perception, history, and biology.

A few credentialed scientists exhibit interest in and research paranormal topics. Sykes' experiences with the Bigfoot community (2016) provides an illustration of the nuances and pitfalls that academic scientists fail to recognize in amateur researchers. While the scientific community generally berates those members who venture into these fringe fields, they view contributions from entirely outside their community even more suspiciously. The amateur will have great difficulty presenting their views in academic circles where a conferred PhD degree is a prerequisite. By default, outsiders' views are considered inferior and may be ignored entirely (Beveridge 1957), especially if they relate to topics that are not part of the accepted sphere of science (Marks 1986). Amateurs face high hurdles to be taken seriously. That does not stop many from trying. ARIGs move their ideas in public, directly and willingly interacting with people who have strange experiences and appealing to their communities and local media. They readily seek out connections to individuals and families who have had their strange story publicized by the media. Many live the mantra "We're here to help," a welcome tone to people who have a need to unburden themselves with what they think may be supernatural stories. A disparate group, ARIGs are not organized under any ideals or principles or societies (like scientists or tradespeople). They maintain a culture of beliefs, not a body of established knowledge. Their subject is "real" because many people experience it. Unlike scientific research, these fields of study are not necessarily about objective truths about nature.

ARIGs are like scientists in that they have a passion for finding evidence (Jenzen & Munt 2013) but starkly different from scientists in their quest for "proof" and the need to "prove" that their idea about what is happening is as they say it is—there are actual ghosts (remnants of the past), Bigfoot, and mysterious sky craft. Absolutes and secure knowledge are comforting. Science, on the other hand, can frustrate with its uncertainty, probability, and tentative conclusions. An audience may be more amenable to a presentation from ARIGs that states they have "absolute proof" even though they have not established this in a scientific sense. Many ARIGs frame their evidence as suitable or even overwhelming for a court of law. Science findings aren't like legal findings. Science deals in probabilities, not absolutes. Scientists strive to conclude that a certain view is "well established" or "highly likely" but we can never have definite truths about conclusions because we don't know everything. The use of the word "proof" (other than in mathematics) acts as rhetoric suggesting strength of the evidence. It's not unreasonable to say I have proof that something weird happened here, and cite the outcomes that support that statement, but anyone who says they have "proof" of ghosts or Bigfoot must be prepared to put on a parade of high quality, robust evidence. Even then, they need to be prepared for an onslaught of nitpicking. The strength of the claim depends on what it looks like after going through a scientific gauntlet. Proof is not for one person or group to declare as definitive.

The Scientific Ethos

Though there is no cookbook-like, easily definable "scientific method," scientists do subscribe to fundamental ideas and well-established methodologies that define how they work and the reliability of knowledge produced. Robert Merton (1942) wrote on the scientific "ethos" as defined by ideals or norms in behavior that scientists feel bound to and that makes science a unique way of knowing (Ziman 2000). Components of Merton's ethos consisted of: communalism (or communism, which is often misinterpreted), universalism, disinterestedness, and skepticism. When unpacked, these practices are revealed to be sensible rather than anything special to science.

Communalism means that the knowledge and the supporting data are *shared*. Science produces public knowledge. Scientists provide sufficient information about what they did and how they did it so that others can attempt to reproduce or falsify the work to make sure it is correct. This tenet also requires that scientific knowledge is archived and organized for others to access. Secrecy and obscurity makes scientific work useless. Only a communal effort can reveal biases and mistakes as well as provide confirmation of results. Originality is stressed so that work is not duplicated. This requires that the researcher be fully aware of what others have already found. Science is very much a community effort, building upon the work of others and not at all like the lone researcher seeking that "Eureka" moment. Because of the interconnectedness of the scientific community, strong objections are inevitable when findings from outside the community of science or outside traditional channels are touted (Ziman 2000).

Universalism represents the ideal where the social context of knowledge is not important; no one authority can dictate what is acceptable. *No one observer is privileged over another.* The observations should be obtainable by all under the same situation. This can be contrasted with "revelation" or intuition where only one person has access to information, or being the "chosen one" that can only see perceive something. Science must be for anyone with the same setup to do.

Disinterestedness is the state of being *not overly invested in the outcome*. Financial, professional or even emotional bias can too easily lead to poor or useless data. Most researchers will assert, "I was completely objective and unbiased during this research." But, bias works in subtle ways that are hard to recognize in ourselves. This bias infiltrates our experiments, observations, and conclusion in major ways. Scientific protocols aim to eliminate that subjective bias. Complete elimination of bias is impossible but many processes can be used to minimize egregious bias that could prejudice one outcome over another. Another outcome of disinterestedness is the reliance on referencing others' work, not your own, to show you are not promoting your own agenda but supporting your conclusions with the body of public knowledge. Disinterestedness caused the unfortunate habit of scientific publications being written in difficult-to-read objective, passive tense.

Skepticism, as a core practice of the scientific community, is the ability to *weigh the evidence in response to the claim being made*. This is reflected in the process of peer review where the work is presented for fair criticism and professional debate. It is a critical component to judge the validity of the research and to advance intellectual progress. All scientists must get used to inevitable criticism and learn to handle the response appropriately.

We can also add *originality* to this list, which requires that you make an effort to know what has already been done and build upon that. Imagination is required for progress, within reason. Effort should be put into finding out something new or enhancing existing information, not running over the same ground, which is an intellectual and economic waste.

The reality of scientific work means adhering to a commitment to testing, checking and disclosure (Haack 2003)—presenting your work openly, being subject to constant criticism, and facing rejection from peers and anonymous reviewers. Your work is expected to make sense, integrating with how the rest of the world works and its natural laws. If the work contrasts against an accepted model of nature, it will face vehement objections. If it doesn't fit with the processes inherent in the scientific ethos, it will be summarily and handily rejected (Pigliucci 2010).

This quote from Dolby (1975) is illustrative of how scientific knowledge develops:

> Every science develops by piece-meal changes, each of which has to be presented in such a way that it is made acceptable to experts who work within the prevailing system of understanding. For routine work, with few implications, all that is really necessary is that the scientist should show that he understand the literature leading up to his own contribution, and that he has carried out his own research competently (in so far as this can be shown in words). More unexpected innovations must be presented with a fuller argument which gives greater plausibility to the novel claims made, and which shows that the new understanding they provide presents clear advantages over the old viewpoint.

Good intentions and valiant efforts are not enough to produce solid conclusions. That depends on adherence to a rigorous (and sometimes unpleasant) process and rigid framework, justifying what has been concluded and correlating it to existing knowledge. ARIGs often are unaware of or reject the tenets of science, but they keep the shallow appearances of science work to maintain credibility.

Image of Science

Science was institutionalized by the end of the 19th century (Daniels 1971). Then, scientists could attain prestige as elite researchers and professionals in society (National Science Foundation 2009). The growing sophistication and increasingly demanding requirements of formalized science meant that the public began to recognize science as an authority where non-experts and amateurs were excluded (Thurs 2007). This exclusionary nature of science is an important characteristic in the discussion of ARIGs.

When the non-scientist thinks about science, what comes to mind? National surveys (National Science Foundation 2009) show that people associate science with three characteristics—having a systematic method, taking place in a special location (a university or a lab), and, to a lesser degree, obtaining knowledge that is in accordance with "common sense" and tradition (Gauchat 2010). The community of science, viewed by laypersons, is characterized by symbolic paraphernalia (white lab coats, test tubes, electronics) and the end products of study (wordy reports, graphs, and displays) (Toumey 1996). The scientific community is also almost exclusively associated with white males, especially historically famous figures like Galileo, Newton, Darwin, and Einstein. We get an over-simplified and over-optimistic representation of how scientists work by depictions from television, movies, comics and literature (Pigliucci 2010) which use these tropes and symbols to reinforce the producer's ideals of science.

In the post-war 1950s, science popularity was high. Pop-culture scientists pushed buttons, used gadgets, and consulted computers. They utilized technology they created. Science appeals to an audience that also has similar higher education and income levels.

The public has little opportunity to see inside the day-to-day workings of a lab, to go on field assignments to collect data, to write grant requests, or submit a manuscript to a journal. Scientists don't typically meet patients, customers, or clients like other professionals. They collect data alone or with colleagues. With publicity about the project held until the end products or conclusions are reached, the public receives no information about the rigorous process undertaken to get there. The press writes little about the actual nature of the research, emphasizing results and value instead. The media presents science as a condensed series of dramatic events, sometimes with premature enthusiasm or high expectations. How often have you seen science portrayed in the media as a years-long effort in collection of data, preceded by writing grant applications and proceeded by the effort of drafting and getting a final paper published? Science is a long, tedious effort to get to a result, but the public rarely, if ever, sees that gradual nature (Nelkin 1987).

Science and the involvement of scientists is used in various ways to lend confidence and authority to an activity or viewpoint (Agin 2006; Thurs

2007). Toumey (1996) refers to "Old Testament style" respect without comprehension. "Scientific" is used as a label of honor, a term of praise that carries the characteristic of being "strong, reliable, and good" (Haack 2003). This "honorific usage" of science is common in our society and reveals a ubiquitous problem—inappropriate mimicry of science. Not everything or everyone who claims to be scientific is so. The manner, language and procedure of science are frequently imitated to appear technical, specialized, and credible (Degele 2005). The "magic stamp of 'science'" (Daniels 1971) has been utilized since the invention of the scientific process. Charlatans pepper their ads and promotions with the word "scientific" to sell products. Thurs (2007) states science was used as an "incantation" in early America; that it was synonymous with the engine of progress. Speaking "scientese" is an easy-to-spot ploy used by advertisers who may be making claims without substantive empirical evidence to support it, appropriating the credibility of science with "high falutin'" big words that sound impressive (Haard et al. 2004). The public takes these as cues of sophistication and expertise of the source suggestive of knowledge and reliability. Advertisers appeal to these consumer mental shortcuts regularly today with blatant use of scientific jargon and images to sell everything from food to shampoo, medicine to makeup (Dodds et al. 2008). Use of science can be an effective marketing strategy (Pitrelli et al. 2006). Since the population viewing these advertisements is likely to not have a high degree of science literacy, they fall prey to a science deception that perpetuates confusion and misunderstanding. How can the average non-science-trained person determine science from sham? It's tricky business to determine if a claim was investigated through an actual scientific method or through a slap-dash imitation process. It is too easy to hijack representations of science. When non-experts play pretend science, it warps science's unique worth as the most reliable way of knowing (Toumey 1996).

The public views science as consisting of textbook facts and absolute laws. Instead, as a body of knowledge, science is always changing and is, thus, difficult to keep up with. We are refining ideas or discarding a disproven one all the time. Scientists themselves are seen as an "other"—technical, practical, analytical, less emotive, even unfathomable (Michael 1992)—guardians of the facts and laws. They are thought to be serious, distant, preoccupied, nerdy, and often not highly socially skilled—an unfortunate and limiting stereotype. Science is a specialty that does not require its practitioners to be trained specifically to interact with the public who do not have a basic foundation in their field. Potentially, those drawn to science as a career prefer less public exchange and their job requirements may not call for it.

Science Talk

Scientists have their own language dependent on their area of expertise and use language in a very precise way—to communicate efficiently and directly with each other. Scientific words are defined with utmost care so nothing is assumed or misunderstood. Different areas of science have their own jargon, which serves a useful purpose to allow for precise communication of ideas between specialists. It saves time and provides important detail but it is truly impenetrable to the non-specialist and even to other scientists outside that field. The biologists will probably not be able to converse in scientific language with the physicist and the anthropologists will not understand the geologist very well. This jargon is spoken in journal articles, at meetings, presentations, and conferences. To unpack and understand scientific language, you may need a PhD degree in that subject area (Mooney & Kirshenbaum 2009). Formal scientific language is "very unnatural" to the outsider (Ziman 2000). Attempts to imitate it are obvious to those with scientific experience and a pretender is quickly spotted. Unlike our typical conversations with friends, family, and even the random stranger in the checkout line, emotion and opinion are not the basis of scientific discussion, the arguments are more sophisticated and concise with inclusions of references to the keys of scientific ethos and norms. It's detached and difficult speech. A hallmark of scientific jargon is the use of qualifying words such as "likely" and "suggests" when discussing conclusions. These words are annoying to the reporter and official who are looking for unequivocal facts. But scientific knowledge consists of probability arguments, not absolutes (Ziman 2000) so the qualifying words have important meaning. The tendency towards probability talk and stating the *best* answer, instead of *the* answer, makes science talk frustrating for some listeners and demonstrates one stark difference between science and other fields such as law, politics, and religion. Science talk contributes to the enforced boundaries around what is and isn't science.

Sounds Sciencey

Science boundaries are necessary to support its exclusivity and authority. Scientists themselves created and actively maintain these boundaries. The exclusivity is vital to the value of science.

Science is a craft with unique skills that must be learned. It can't be done by just anyone who deems to call himself "a scientist." This is not dissimilar to becoming a lawyer or a medical doctor, or entering the clergy, as certain requirements are imposed upon the individual to follow. The science community requires that a thorough understanding of how science works must come from *doing* science including undergoing graduate-level specialization, publishing in scientific journals, obtaining membership in professional organizations, and accepting and using the shared practices and concepts of the community. Qualifications as a scientist cannot be achieved through absorption or observation (Pigliucci & Boudry 2013), only through participation in the structure itself.

The power and privilege of science in society resulted in the creation of a corrupted *doppleganger*, where science is mimicked or co-opted by non-scientists or pseudoscientists to suggest the individual is inside the boundary of science and to gain a pretense of authority. Their activity *sounds sciencey* but it lacks the substance of science. Popular physicist Richard Feynman described this imitation of science as "Cargo Cult science" in 1974.[2] The term was derived from "cargo cults" a phenomenon that arose with natives in Melanesia (the South Pacific) during World War II. Traces of these "cults" may even be found today.[3] Feynman used the term "cargo cult" because he noticed the same type of activity in a different context. Here is his description of cargo cults and how he related it to pseudoscience or "science that isn't science":

> In the South Seas there is a cargo cult of people. During the war they saw airplanes with lots of good materials, and they want the same thing to happen now. So they've arranged to make things like runways, to put fires along the sides of the runways, to make a wooden hut for a man to sit in, with two wooden pieces on his head to headphones and bars of bamboo sticking out like antennas—he's the controller—and they wait for the airplanes to land. They're doing everything right. The form is perfect. It looks exactly the way it looked before. But it doesn't work. No airplanes land. So I call these things cargo cult science, because they follow all the apparent precepts and forms of scientific investigation, but they're missing something essential, because the planes don't land.

Modern ghost hunting may be the best example of "cargo cult science" ever as the participants follow the leader and use re-purposed equipment, playing the part of a scientist, but missing the essence of real science which is far more complex and involved—a practice outside their scope of life experience. Explaining cargo cult science to someone requires a backstory. A suitable term doesn't exist to describe people sincerely going through the motions of science with a belief that the outcome will be equivalent. I proposed the self-explanatory term "sounds sciencey" to describe activities and demonstrations that attempt to superficially portray science and its associated credibility but lack the meaningful substance of it. Behaviors indicative of *sounding sciencey* include clothing oneself in the metaphorical, and sometimes literal, garb of science (white lab coats and gadgets, for example) and attempting to speak the language. Bader, Mencken and Baker (2010) discusses a unique vocabulary that develops in subcultures called *argot*. This specialized talk serves to distinguish insiders from outsiders. Such vocabulary is common to those involved in many hobbies such as hunting, craft-making, role-playing games and, as we clearly see here, paranormal investigation. Some ARIG argot examples include "P-G film," "EVPs," "implants" (from aliens), and "orbs." Hundreds of examples are found repeated over the extent of ARIG-related media. This special language is not the same as scientific jargon, however, as scientific words are formally and precisely defined in journals and cited thereafter, used as a shortcut for very long technical descriptions. Argot is more fluid with the different purpose of flagging someone who isn't a legitimate part of the subculture or a "noob"—short for "newbie," a person new to the situation or subject without significant experience.

Another ploy to gain credibility is to invent an organization with an official-sounding name and acronym. The public is accustomed to these acronyms from academic professionals, and military organizational affiliations, so this is an effective means to gain credibility.

ARIGs frequently use sciencey-sounding words and actions with a public audience that only a genuine expert in a specialized professional field would recognize as incorrect or misleading. When ARIGs talk about their investigation in the guise of a scientific investigator, being *scientifical* (falling short of being scientific), actual scientists in the audience notice the mistakes like a musician would notice notes played off-key. Dr. Bryan Sykes noted that cryptozoologists who talk about DNA samples are embarrassingly mistaken because they clearly do not grasp the nuances and complexity of the field (Sykes 2016). Ghost hunters who discuss quantum mechanics and "ghost science" trip up on basic definitions and established concepts. Ufologists fail to reconcile their ideas with physical laws. Use of sciencey-sounding words often betrays lack of deep understanding and is invoked for appearances. Such play is not new; it was noticed long ago by professional spirit seeker of the Society of Psychical Research, William Crookes, in 1870. Playing pretend scientist is not a suitable substitute for being accurate and having sound and reliable claims of knowledge. It's a hollow shell covered with skin in comparison to a complicated system of organs and strong supportive framework that makes up a living, working body. If you look too quickly, it may pass for real, but it does not function like the real thing.

8

Science and the Paranormal

"All argument is against it; but all belief is for it"—Samuel Johnson[1]

Science Rejects the Paranormal

Long before the concept of science and scientists existed, belief in the supernatural was the norm. People have reported ghosts and monsters for millennia. The advent of scientific thinking and technology reduced supernatural and superstitious thinking, but did not eliminate it entirely. Media continues today to spread and enhance belief in supernatural ideas that some portions of the population retain. The paranormal is even more palatable and believable than supernatural agents. When presented by people that we consider authorities (whether that be scientists, those who approximate scientists, self-proclaimed experts, or media personalities), belief in paranormal or fringe concepts can sound highly plausible and they enter normal conversation streams.

Mainstream science—academia, credentialed scientists and science communicators—has considered, but ultimately rejected, various paranormal phenomena as genuine. As I described earlier, there are many examples of organized scientific research into seemingly paranormal subject areas such as parapsychology, hauntings, unidentified aerial phenomenon, and unknown creatures. The paranormal was the subject of serious academic inquiry in the late 1880s (Stoeber & Meynell 1996). Psychical research was undertaken for 100 years and is well-documented. Journals existed and societies were staffed with learned men. Parapsychology, as a field of scientific standing with the American Association for the Advancement of Science, diminished. A watered-down version became the "everyperson's science" of paranormal investigation. Even after a concerted effort to produce good evidence, there still are no reliable facts to support parapsychological phenomenon (Baker & Nickell 1992; Gibson et al. 2009) and a consensus was never reached (Irwin 1989; Stoeber & Meynell 1996). Paranormal encounters can cause intense fear and anxiety, yet traditional psychology journals treat parapsychological topics as taboo (Houran & Lange 2001). UFOs, once of interest to renowned scientists like Carl Sagan, lost all credibility after several formal examinations of the evidence by scientists and the military (Sagan & Page 1972). A well-known paranormal writer, John Keel (1975) admitted that rational people eventually left ufology, leaving it to "cranks, publicity seekers and paranoics." In 1992, the prestigious Massachusetts Institute of Technology (MIT) held a serious conference about alien abduction because interest in the subject at the time was so high (Bader et al. 2010). Cryptozoology was once tenuously professionalized under the International Society of Cryptozoology. Once it dissolved, with no credible organizations promoting any standards, the field went more commercial. Hoaxing abounds by those who claim to be legitimate researchers. A few academic scientists took cryptozoology seriously at their professional peril. The search for anomalous primates took a turn for the worse when amateur enthusiasts and even some scientists made hasty pronouncements. Today, reputable scientists are loathed to pursue claims due to the high risk they will be hoaxed (Regal 2011). So we see why the study of non-material concepts is not a fruitful path for scientists (Lyons 2009) and, therefore, we have a split community of those who examine paranormal claims from an objective perspective, and the amateurs who are invested in the belief that there is something paranormal awaiting discovery.

The relationship between the orthodox scientific community and the paranormal community is a messy one. To use a metaphor, imagine two societies on different islands, evolving independently, and only rarely accepting news or information from the far away island. The natives of each island have their own leaders, languages and customs, their own artifacts, relics, and belief systems. Hostility may rise towards "the other" side. ARIGs regularly assert there has not been a concerted effort to study these things, not enough money is available, or scientists are closed-minded and stuck in the conservative paradigm. The history of research by scientists demolishes this argument. Bryan Sykes discusses this point most recently in his book about cooperating with Bigfoot and Yeti enthusiasts (Sykes 2016). He set out to show that he was taking a scientific look at the subject and that complaints from claimants that they were being ignored was unfair. In his situation, Sykes had success in forging a relationship of trust between these disparate communities. Scientists aren't easily impressed by anecdotes or stories passed around town. Sykes set out to look for DNA, a strong form of physical evidence. Scientists would be eager to examine body parts or remains, or a collection of carefully measured and documented anomalies that could form the basis of a focused search. In the early stages of scientists investigating ghosts, cryptids, and UFOs (among other anomalies), speculation was rampant and many fringe claims seem plausible. Several creative explanations are considered. These ideas fall away if no evidence turns up to support them. If there was something interesting worth looking at, it has been examined and found lacking for good reasons. Paranormal or fringe explanations do not match with well-established knowledge about nature causing scientists to reject these explanations. Cryptozoology, for example, conflicts with not only biological and zoological tenets but also clashes with strongly supported theoretical frameworks in geology and paleontology. Attempts to replace an accepted and tested idea in science with a speculative and strange new one will not stick. Evidence (and plausibility) in support of the new idea must overwhelm the old to allow for a massive shift in acceptance. Thus, extraordinary claims *must* be supported by extraordinary evidence (Loxton & Prothero 2013).

Excluded from the Scientific Establishment

When subject areas leave the scientific realm, they take on a life of their own in the popular culture. ARIG subject areas remained very popular with the public even as they diminished in scientific respect.

Being a member of any elite community is a draw and amateurs will seek to be part of a group. With acceptance into the academic professional realm not typically open to ARIGs, they appealed directly to the public as having elite knowledge in these fringe areas and called themselves "experts." This, however, meant that the quality of research was not equivalent to a scientific level and setback the reliability of knowledge. Clarke (2012) in *Natural History of Ghosts* says that the takeover of paranormal subjects by untrained enthusiasts undid much of the past serious and respectable research into the subject areas (p. 283).

Animosity and Resentment

There is certainly an element of class distinction between the professionals and the non-professionals. In defense of their efforts, ARIG rhetoric portrays scientists as afraid of new ideas and seeking to protect their status and livelihood. They describe orthodox science as done by "close-minded eggheads" in an "ivory tower." This understandable attitude towards monolithic "science" as being closed off perpetuates the communicative distance and lack of exchange between ARIGs and the scientific community. ARIGs may also display a desire for revenge against the scientific establishment that rejects or ignores their claims. They hold that they will be the ones to break through and make the ultimate discovery, not scientists (Loxton & Prothero 2013). But as documented, many scientists had and still have excitement for discovery of strange new things and would be delighted to be the ones who help bring a sensational and real discovery to the eyes of the world. New information in nature does not pose any threat or undermine the scientific enterprise. While a few scientists are outwardly sympathetic to paranormal explanations, individuals with a scientific title or career are only rarely explicitly identified with ARIGs or members of such a group. Scientists will focus on research that can be immediately useful, seeing fringe subjects unworthy of attention, even to a degree that the participants are openly mocked for wasting their time, leading to open animosity towards scientific or skeptical involvement in these now-democratized fields.

Attention from science helps legitimize the subject. Paranormal proponents *want* the luster science can bring to their area of interest, but they hate that they can't get it. When science ignores, rejects, or shows disdain towards a subject area that is perceived as legitimate by researchers of that area, it breeds mistrust, suspicion, and feelings of contempt. Scientists, with their books and journals and pronouncements, are not wanted unless they are aiding your cause. Scientific or expert critical opinion is shut out unless it can be used to reinforce a preferred position. Some ARIGs reject any intellectual approach, asserting that the real discovery takes place through active field work, not via "armchair research." Paranormal skeptics who wish to contribute to the discussion or ask questions are excluded from participation in events, banned from Internet forums, or ridiculed in pro-paranormal publications. Another example of the hostility exhibited by amateurs in paranormal research was the response of the cryptozoological community to three books published from 2011 to 2016 by science-minded and skeptical scholars. All three brought to light significant and critical information about mystery animals that should influence further pertinent research. Yet, they were deliberately ignored or given negative reviews by pro-cryptid commenters. In 2011, *Tracking the Chupacabra* by Benjamin Radford neatly solved the mystery of the *chupacabra* as a complex cultural beast reflecting community fears in South America that spread to the U.S. mainland and around the world via the Internet. Loxton and Prothero's *Abominable Science* (2013) dug into the history of many cryptids to discover how some of the key people and tales that were highly regarded were baseless and should not be regarded as quality evidence. One Bigfoot extremist even called for the publisher to retract this book. Several reviewers pilloried the book while admitting they didn't read it. The latest, *Hunting Monsters*, a scientific comparison by Naish (2016), followed the pattern now expected by the split crypto-community (that of the amateur enthusiasts versus scholarly researchers who value application of the scientific ethos). If a book exposes evidence against the biological reality of cryptids, no matter if it provides worthwhile social and historical information, it is rejected. Disregarding and attacking sound, referenced arguments from critics and legitimate experts is an indication that a field is pathological and cannot progress in knowledge.

Public Interest Remains

Even though the scientific community rejects an area of research as worthless, the public is undeterred in interest. To them, these topics are still "mysterious" and deserve attention. "Isn't that what scientists do?" they ask, "Investigate the unknown?" People still report UFOs, ghosts, and Bigfoot and expect investigation of these mysteries by someone. Understandably, the public wants science to take a more serious interest in their favorite fringe subject, but they are unaware of the long history and issues I described above that limit modern scientific investment. Professional scientists focus on solvable problems and abandon dead ends. To dwell in fringe areas would be detrimental to their career. The evidence for paranormal ideas, unfortunately, failed to be extraordinary, so the scientists turned away (Marks 1986). As academic interest dwindled, the investigation and research on these topics shifted. Anyone who put in time and effort, without the requirement for higher levels of education, was free to establish himself as a researcher, writing or expounding on the subject—the amateur claimed the ground. With direct access to witnesses and the media, the fields evolved into a different niche that wasn't science and, in many circumstances, rejects science outright. Views were diverted from naturalistic explanations as required by science to non-scientific speculation that resulted in recognition in society, media attention and, sometimes, even profits.

If new evidence came to light regarding ARIGs' areas of research, scientific experts would be all over it, jumping at the opportunity to take over from well-meaning, enthusiastic amateurs and a formal scientific process of meticulous collection, analysis, and publication would be enacted. If an amateur cryptozoologist caught Bigfoot, within days the study of Sasquatch would no longer be the realm of amateurs. ARIGs realize this, albeit somewhat grudgingly. At their point of highest success, it is a galling thought that they should have to turn to professional scientists who will vet, publish, and eventually attach their own names to any worthwhile findings. It seems unlikely that those invested in the pursuit would wipe their hands, say "Mission Accomplished," and walk away. There is too much invested, emotionally, in this very involved hobby (Regal 2011). Ideally, scientists should graciously recognize contributions from amateurs as warranted, provide advice and opinions when requested (or when opportunities arise). Scientists would benefit from attempts to understand the amateurs' goals and role with the public. However, scientists work long and hard, collectively and individually, to earn their place in the hierarchy of the knowledge process. They are the gatekeepers for what is in scientific interest. Amateurs can not gain that rank. Their contribution always remain one step removed.

Higher Learning

Several unaccredited universities and parapsychological institutes play the role of places of higher learning for fringe topics. Examples include the American Institute of Metaphysics,[2] the Rhine Research Center,[3] the Nevada Institute of Paranormal Studies,[4] Flamel College,[5] and the International Metaphysical University.[6] A few religious-themed ministries offer an Internet course in demonology and exorcism (Krulos 2015). Obtaining credentials from such places may look impressive to some, but they hold little gravitas as they are not accredited as educational sources. There are no restrictions against offering such classes or certifications. There are not genuine experts in cryptids, ghosts, UFOs, or demonology because they have not been established as valid phenomena and, thus, there is no basis for their study. Without any standards and foundation, anyone can decide to call themselves an expert in a field that they contrive. Therefore, we see the low bar for participation in these fields. Self-directed learning is the norm. Unfortunately, not many ARIG members dig into the deep history of their fields to assess what has been done before. Use of the prior literature on a subject is a yardstick used by scientists to judge how competent a researcher is (Dolby 1975). Many ARIGs presenting to the public show superficial or incompetent knowledge of the long history of their these subjects, focusing only on the modern media era. Even today's top paracelebs will be found woefully lacking in suitable background knowledge. My favorite shortcut to testing the knowledge of self-described expert ghost hunters is to ask about the work of the "SPR" and see if they have even heard of it.

Personal Values

ARIG participants express their personal value systems in their activities. Science should ideally by "valueless" in the sense that no particular value, except knowledge, guides the outcome. When values become enmeshed in scientific discussions, those discussions often become unresolvable and lead to animosity. A perfect example of a values-laden social issue that can't be resolved is that of abortion. No amount of scientific facts will be able to resolve issues that are tied irrevocably to personal values and beliefs. The influence of values is also evident in the context of science and the paranormal.

Early in the history of professional science, some scientists rejected the idea of materialism—the tenet that says all phenomena, including that of the mind, can be attributed to physical aspects of nature—and felt that a human was more than just matter. He also had a soul, conscience, or moral sense. When scientists studied psychic phenomena to find evidence of life after death, they insisted that there were forces yet to be discovered. Even at this time, some wanted the boundaries of science extended to include "supernatural" or non-material observations (Lyons 2009). The scientific community rejected this deviation and the rigid framework was retained. The same call to soften this structure is still heard today with allegations, especially from parapsychologists, that some phenomena are beyond the scope of reductionist science, and that it is folly to reduce everything to natural laws, matter, and mathematics.

Alternative ideas about the natural world that were less than scientific became popular in the 1960s. These new fields, as alternatives to orthodox science explanations, were a means to connect to personal values systems (Thurs 2007). When scientific methods did not provide the answers that were sought, proponents found other ways of framing the subject, frequently venturing into supernatural explanations (Denzler 2003). Examples of this are abundant in New Age spirituality—astrology, astral projection, spiritual enlightenment, expansion of consciousness. Blended ideas of natural laws and spiritual suggestions permeated the UFO community in particular. Feel-good spirituality also seeped into concepts about hauntings and mystery creatures where psychic and trans-dimensional themes emerged. Aliens became our space brothers trying to warn us of coming disasters. Bigfoot became a person of the forest connected to nature, like humans used to be long ago. While this volume is not an appropriate place to address this complex subtopic, there *is* much to consider about the spiritual aspect of those who may replace traditional religious practices with paranormal beliefs. (Consider that the basis of most religions is supernatural, but we typically don't think of religion as institutionalized supernatural belief.) Most people are invested in a belief system that gives their lives special meaning. They may be seeking an answer for some dissatisfaction in life. Some scholars suggest paranormal beliefs are adopted in response to a rejection or resistance to orthodox religious practice or rigid scientific materialism. Others say people turn towards fringe and occult ideas because they are disenchanted by science and a scientific world view. Narrative power is placed back in the hands of the individual instead of some institution (Jenzen & Munt 2013; Bader et al. 2010). At the intersection of leisure time, paranormal beliefs, and a search for meaning in life, the environment of "occulture" grew. Occult-related cultural items blend, disseminate, and became influential in society (Partridge 2013).

It is useful to examine one time in American history when values held by the scientific community, the public, and an individual promoting an alternative to science clashed in a pop culture explosion. Not many ARIGs would know of the Velikovsky affair; they might find it unimportant and irrelevant, but the scenario and reaction that occurred is illustrative of the reaction to ARIGs, their subject areas, and their methods.

The Velikovsky Affair, or How to Tell What Is Worthwhile Scientific Work

A fascinating discussion by R.G.A. Dolby (1975) provides a guide to a case study about a popular idea that was nearly universally rejected by orthodox scientists, sold directly and successfully to the public by a non-expert, and even involved religious connections. Immanuel Velikovsky wrote in his 1950 book *Worlds in Collision* that a comet from Jupiter caused catastrophic disasters on Earth and Mars and eventually became the planet Venus. These disasters could be correlated to events in the Old Testament of the Bible. Velikovsky was not a scientist and provided a physical mechanism that was nonsensical to those knowledgeable in basic facts of nature. He used historical documentation and speculation instead of astronomical calculations or other quantifiable methods. Scientists didn't take him seriously, rejecting this absurd tale out of hand. But Velikovsky supporters saw this new proposal as revolutionary science and the affair resulted in undermining the authority of scientific experts in popular culture. Dolby describes the rise of Velikovsky's popularity and community conflict as an example of the social nature of science. There are so many individuals who present revolutionary ideas—how do we know which ones to take seriously? In a specialty scientific field, a contributor is expected to present their arguments and supporting evidence in a formal paper to a reputable journal and be subject to peer review. If the paper does get published, the idea is deemed worthy of consideration by this community. The proposed concept is expected to fit into the existing understanding of the field or improve upon what is known, citing literature of what has come before. Velikovsky's proposal of catastrophism did not go this route. His unsupported theory spanned disparate fields of study. However, the literature cited regarding physical sciences revealed to Dolby that Velikovsky had no understanding of the foundations of the field.

There were considerable objective reasons for the scientific community to reject the Velikovsky idea. The proposed theory, while explaining some anomalies, was too selective in scope and created more and greater anomalies than it solved; it did not address why the alternatives were not suitable and was not defendable with evidence or by argument. Velikovsky bypassed peer review, appealing directly to the public with a dramatic book. This created a subjective reason to reject it—it short-circuited and undermined scientific process and authority. His idea, thought ludicrous by reputable experts, was taken seriously by the public and became fashionable, making his book a blockbuster best-seller (Gordin 2012).

The same scenario applies to ARIG theories. Ideas about alien visitation, Bigfoot, hauntings or demonic interference sound bizarre to scientific experts. But they become popular and the subject of media exposure because they serve a need in our culture. Ideas about paranormal activity also span disciplines, adding to the difficulty of fitting it into a scientific format. As proposed, spirit activities, anomalous aerial phenomena and trans-dimensional "zooform" (animal-like) creatures are challenging to explain in terms of modern scientific knowledge—they would be revolutionary discoveries that would change the current mode of thinking about the world. Looking back upon the history of psychical research, parapsychology, ufology and cryptozoology, they have not followed the path of a revolutionary shift in understanding. As Dolby describes, a revolutionary idea would result in a new stand-alone discipline that would eventually develop structure to generate and test hypotheses and, in time, progress in understanding. This new field would draw in researchers, it would flourish, gain respect, and generate new knowledge that would tie into existing knowledge to supply an overall cohesive view. It would make sense. Parapsychology was on its way along this path but stumbled and not progressed. Other fringe fields remain publicly popular but have also not generated reliable knowledge and have stagnated. This argument supports abandonment or wholesale change in pursuit of these fields. However, these fields and ARIGs' pursuit of them have social value, so they continue to exist incommensurate with scientific study. They remain separate from scientific fields and processes and pursue their specific goals on their own terms with different language, methods and assumptions.

Mind the Gap

As with Immanuel Velikovsky's idea about worlds in collision and catastrophic events in the solar system, ARIGs' ideas about life after death, visitors transporting across our skies, and mystery animals hiding in the shadows resonate with many people. They are popular concepts regardless if they make scientific or logical sense. Normal people, not academics, are reporting these events and expressing the desire that they be investigated. Who can they report them to if not the scientific authorities? ARIGs provide considerable social function by serving as a source for those who have witnessed these events, other interested amateurs, and the media (Westrum 1977; Hynek 1972).

The field of ufology is the product of social effort, not that of a cadre of intellectual elite (Blake 1979). Writer John Keel (1975), who was influential in UFO culture, described the UFO community at its peak as consisting of nonprofessional, nonsocial people with identity issues and a lack of higher education where active participation provided an ego trip and an escape from an "undistinguished life"—a harsh assessment but somewhat compatible with what I found with ARIGs as well. Because grassroots UFO groups did not use statistics or similar scientific methodologies to address anomalies, their work was not considered by scientific journals. Therefore, they started their own journals, establishing their own criteria for publication that was much lower than that of standard professional journals. UFO groups eventually became outwardly antagonistic towards each other and developed personality cults. It is fair to say the same thing is occurring with other ARIG areas such as ghost hunting and cryptozoology teams as certain high-profile people wish to gain a position as "*the*" top expert. Ufologists did it first. As with psychical research in the late 1800s, the mid–1900s was a time when credentialed investigators affiliated themselves with UFO organizations that addressed certain questions that science avoided or discarded ––reports of alien visitations, missing time experiences, abduction scenarios. It was an unexplored area, and perhaps these academic investigators presumed there was something to be discovered. The government, universities, and individual academics eventually retreated and only a few popular figures at least tangentially affiliated with science remained associated with ufology, which was now heavily weighted in discussion of alien visitation. Budd Hopkins, John Mack and Stanton Friedman wrote books and were treated as experts because of their credentials but they were ostracized by the scientific community. Instead, they appealed directly to the UFO advocates and garnered support for their questionable ideas. A similar pattern is seen in both ghost and cryptozoology research. Not many academics engage in serious ghost hunting but some have moved into the field of anomalistic psychology which takes these strange experiences seriously, not in terms of the supernatural elements, but in how our brain creates and perceives events we interpret as paranormal. Researchers in this area include Dr. Richard Wiseman and Dr. Christopher French in the U.K. For cryptids, notably Bigfoot, credentialed men such as Dr. Jeffrey Meldrum, Loren Coleman and Dr. Karl Shuker are cryptozoological advocates.

Ghost investigators, most of which whom do not have higher education or professional standing, make their reputations via publicity from TV shows. There is a distinct disconnect between television and television-influenced paranormalists and academics working in parapsychology. This is dramatically demonstrated by consulting parapsychology references like Cardeña et al. (2015) and examining the work of the ASSAP (Association for the Scientific Study of Anomalous Phenomena) (ASSAP), the Koestler Parapsychology Unit (KPU) in the Psychology Department at the University of Edinburgh in the U.K., and the mediumship research at the Windbridge Institute in the U.S. ARIGs do not resemble these examples.

ARIG communities are insular. Their ideas are not subject to criticism or peer review and do not become part of the overall intellectual discourse as scientific efforts would (Regal 2011). Scientists are skeptical of reports from outside their community, in part because of fraud and error, but in the cases of amateurs or even those outside their professional field of expertise, various factors are at play that curtail progress of unorthodox ideas (Westrum 1978).

After many decades of dedicated individuals researching and investigating these topics, in the eyes of the scientific community, no high-quality, reliable evidence exists that would shift the consensus thinking about paranormal phenomenon. Advances in technology and markedly increased interest by media corporations to delve into these areas have resulted in no significant discoveries that would turn the tide of established thinking. Therefore, scientists have adequate justification in ignoring the phenomenon (Marks 1986) as it does not pose any useful research questions or provide any credible data for them to examine. Yet, paranormal events seemingly still occur. That is, witnesses perceive them to be unexplainable and paranormal. When the scientific community doesn't appear interested or they decry the topics as ludicrous, where does the public turn? They go to self-styled experts willing to step in as authorities. ARIGs relish this role.

Bigfoot's DNA

In 2012, a team of professional scientists from the University of Oxford and the Museum of Zoology, Lausanne, led by Oxford geneticist Dr. Bryan Sykes, put out a public call to receive DNA samples from suspected anomalous primates including the Asian Yeti and the North American Bigfoot/Sasquatch. The samples would be prepared and genetically tested, then compared to sequences in the worldwide genetic database GenBank to ascertain with which animal they best aligned. The cost of the analysis was subsidized by a television production company who made a documentary program about the quest for answers to these mysterious sightings (Sykes 2016). The project was a true cooperative effort between amateurs who claimed they could obtain physical evidence for cryptids and experts who could assist them with unprecedented access to the latest genetic technology. A total of 57 samples were received from around the world, mostly from amateur researchers. Of these, 37 (presumably those of highest quality) were selected for genetic analysis. Eighteen were from eight U.S. states. Eight samples were anticipated to be from the *almasty* of Russia. Three samples were collected in the Himalayan region of Asia, and one came from Sumatra, supposedly representing the anomalous "man of the woods" *orang pendek*.

Earlier in 2013, long-awaited results appeared from a similar project. A five-year study done by Ketchum et al. also collected from amateur researchers supposed DNA samples belonging to anomalous primates. While Dr. Sykes was a recognized academic expert in DNA, Ketchum was not. Critics identified several serious problems with the Ketchum study from sample collection, analysis, interpretation, and conclusions, to the scientists themselves. Dr. Ketchum, a veterinarian, lacked qualifications to be considered an expert in this academic area of genetics (S. Hill 2013). She and her team concluded that the results showed Bigfoot exists and is a human hybrid—a human crossed with an unknown primate. This implausible conclusion was disputed by academics who declared that her study was "very poorly executed" and "never gets close to providing the exceptional proofs that such exceptional claims require" as noted by Sykes (2016). The Ketchum study was disregarded as solid evidence for anything while the Sykes et al. study was regarded as the first credible attempt to identify anomalous primate DNA. Sykes' results were published in July 2014 in the

Proceedings of the Royal Academy of Sciences B. Unfortunately, results revealed no anomalous primates in the lot. The sequences all matched 100% with known animals, there were no "unknowns" (Sykes et al. 2014). All the U.S. samples turned out to be from animals already existing in that area—cow, horse, black bear, dog/wolf, sheep, raccoon, porcupine, or deer. The Russian samples were also disappointing, as was that of the *orang pendek*. But two of the Himalayan samples were interesting and generated significant press coverage and positive press for cryptozoology. Samples from Ladakh, India and from Bhutan matched a fossilized genetic sample of an extinct lineage of polar bear, *Ursus martimus*, indicating an ursine lineage interpreted to be unknown in this area. As science typically goes, these interpretations were challenged by several other genetics and zoology experts who argued the samples were within the range of variation of normal bears.

The Sykes study plugged into the heart of cryptozoology and reflected the efforts of ARIG activities. As it is practiced today by amateur Bigfoot hunters and monster trackers, cryptozoology is not done in a scientific way. Yet, this project was an excellent example of amateurs working with professionals, providing the raw material for analysis. Even though the conclusion was not groundbreaking, this teamwork is what is needed to make genuine discoveries (however unexpected) and come up with sound (and maybe someday, surprising) answers for amateur anomaly hunters. Cooperation between professionals, amateurs, and the funding source deserves praise—it served as a template for other such studies.

9

ARIG Portrayal of Science to the Public

Science in the media is often framed as a form of entertainment, not as necessarily critical information. In *Selling Science* (1987), Dorothy Nelkin reminds us that scientists fill a role in press coverage: they appear as "wizards," superior in knowledge and intelligence to the rest of us. The media portrayal of scientists creates a serious misconception of how science works. And, it places scientists on a pedestal of authority. Nelkin also says that press coverage of scientific findings is framed to appeal to the public's desire for an easy solution to problem issues. By not presenting the entire process of scientific inquiry or discovery and by de-emphasizing important points of dispute or possible error, journalists leave readers with little to judge the reasonableness or truth of such findings, or even if they are scientific at all.

Over half of the ARIGs that I researched used the culturally-established authority of "science" as a stamp of legitimacy for themselves. They stated that they were doing *scientific* research or following a *scientific* method. Only someone familiar with science or a scientist himself would carry the insider knowledge to see that ARIGs nearly always fall short of this mark. The following quotes from ARIG websites hint at why these groups appeal to scientific authority to promote themselves:

> ... in order to provide proof of an observation, one must connect it to some "provable" reality.... The result of backing up observations with science enforces reduction of heretical activities and engenders trust with a client[1];
> ... where observations become more powerful than myth[2];
> Our scientific approach makes us one of the most comprehensive and accurate in the field.[3]

Why appeal to scientific authority to promote their group? Because they believe scientific proof is the "best" kind of proof. They understand that scientific results are valued in our society as the high bar in support of a claim. The use of gadgetry and trappings of science look impressive to the client or audience, applying a veneer of whiz-bang legitimacy on to the activity. As one group above noted, equipment readings help to quell the reputation of gullibility and strangeness associated with belief in ghosts. It's not just your imagination, there is something going on here, they conclude. ARIG participants and leaders use the trappings of science to show they deserve to be taken seriously because they were committed to a rigorous method of getting to the truth. The sciencey language and equipment is a key to projecting competence, professionalism, and reliability for many groups. Those that invoke science reveal a strong desire to be seen as more accurate and better qualified than the group on the other side of town or on a TV show. By sounding sciencey, ARIGs attempt to create special boundaries around themselves analogous to how scientists wall-off their territory of expertise from amateurs.

Attempts at establishing legitimacy by appearing to be scientific are observable in ARIGs' emphasis on systematic methods of investigation, use of highly technical and superficially impressive equipment, a set of training processes for members, and certification, affiliation or connections to schools, organizations and institutions. Yet, they have only borrowed the authority of science—some say "conjured" it (Toumey 1996) in the sense that it comes out of nowhere, without foundation. The average ARIG member has no scientific training. Therefore, their formal understanding of scientific research does not vary from that of another layperson. ARIGs perceive science through pop culture and everyday experience, same as the public. Yet, they invoke it and attempt to speak the language and follow the processes they think they should to be scientific in their endeavors. Unfortunately, confusion and muddled thinking regarding what science is and how to do it was apparent in the ARIGs' materials. ARIG member frequently cite topics such as psychology, electricity, geology, zoology, and quantum physics as related to the paranormal, but they only have a cursory knowledge of these complex fields that require several years of dedicated formal education. Individuals conducting genuine scientific research would reasonably be expected to understand critical scientific concepts such as validity, controls, objectivity, bias, interference, statistical analysis, skepticism, and peer review, but these concepts are almost never mentioned in relation to ARIG processes. ARIGs conduct research and investigations with what they believe are sound methods, but, like actors on TV, they are playing a role. That role-playing may convince an audience or each other, but is cannot produce high-quality results that will be taken seriously outside of the paranormal community.

ARIGs will frequently take a certain attitude towards the usefulness of science dependent upon its utility for a certain situation. That is, they will invoke science in one setting to appear credible, but reject it in another setting when the objective truths of science appear to create a hurdle to their preferred paranormal explanation. In deference to their belief in an ultra-normal entity or conclusion, they will depict science as stodgy and closed-minded, blind to the obvious paranormal conclusion. As an underdog versus orthodox science, they assume the role of the hero, to be mavericks, or to state they are doing cutting-edge science. Claiming of a special heroic, rebel role is also an ego boost. They feel they "know" a truth that has not yet been revealed and amateurs can revel in a sense of revenge against the institution of science (Pigliucci 2010).

ARIGs also represent to the public an accessibility that one doesn't have with professional science. It is difficult to corral an astronomer to help investigate your UFO report, and the wildlife officer will probably not take your Bigfoot sighting seriously. But ARIGs are glad to do assume the role of knowledgeable expert who will take our strange claim seriously. Trained scientists will quickly spot flaws in this game. They recognize all the potential pitfalls in relying on eyewitness stories that may produce mistaken or unfounded conclusions. Expert zoologists, geneticists, astronomers, and psychologists see through the ARIG's sciencey language and attempts to sound sophisticated. Sounding sciencey only works well for those without high scientific literacy.

Scientifical in the Media

Where an earlier generation was inspired by *In Search Of...* and *Unsolved Mysteries*, the wave of ARIGs of the 21st century have stated outright that they are inspired by TV shows such as *Ghost Hunters, Ghost Adventures, Monster Quest* and *Paranormal State* (Potts 2004; Brown 2008). Admitting they were influenced by these shows strongly suggests that they believed (at least at one time) what they were viewing had some reasonable basis. ARIGs admit to the practice of taking notes from TV shows and explicitly following their procedures. For a specific example, I note how ghost investigators adopted the term "reveal" as used on *Ghost Hunters* to describe the discussion of evidence with the client. And, some, but not all, Bigfoot hunters have adopted the inelegant term "squatchin'" to describe their efforts seeking Sasquatch which was derived from the show *Finding Bigfoot*.

ARIGs present a view of what it means to do science *based on* what they have gleaned from imagery of all kinds, descriptions in books, magazines, movies, television, and radio—the same as what the non-specialist public views. The mass media relies heavily on a high entertainment factor to maintain audience attention. Therefore, the mass media tells an exciting "story" of science (Gregory & Miller 2000) that does not accurately depict how genuine science works from day to day and long-term.

Walter (2013) looked specifically at faith *plus* scientific aspects of paranormal television shows in the early 2000s. She concluded that the portrayal of investigators, namely those from *Paranormal State*, was a display designed for believers using science and technology to support explanations for entities based on faith. Selected, even contrived, elements of science were adapted to this paranormal conclusion. The acceptance of nonscientific faith and anecdote contrasted markedly with the technology and stated scientific methods utilized. Walters remarks that this scheme goes back to Victorian times, to the first scientists who investigated the unseen world with what was then modern technology of galvanometers and photography. This type of portrayal, blending common beliefs of the public with the appearance of scientific rationalism gives an impression of "common sense" research and conclusions. To those who already believe in the paranormal, such depictions are reinforcing. To those that don't, the strange mix of faith and science is "uncomfortable," an "amalgamation of fact and fiction," producing an "authoritative artificial body of knowledge accepted as true by believers" (Walter 2013).

She provides an excellent example referencing a *Paranormal State* episode where the Paranormal Research Society (PRS) team is helping a woman named Lara who supposedly is possessed by a demon:

> Because Lara's case is so severe, however, the PRS team determines to use experimental "scientific" means in order to aid the religious efforts of all involved. To this end, Ryan [Buell willingly subjects himself to the Ganzfeld Experiment, in which ping-pong balls are taped over his eyes and noise-cancelling headphones pump static into his ears; as he stares into a red light, he is meant to experience a sort of blindness. Through this sensory deprivation [parapsychologists] believe that the subject lies often to extrasensory stimuli and might be able to see entities and ear voices from the invisible world. In addition to the Ganzfeld apparatus, Ryan dons the Shakti Helmet, a device developed by neurotheologist Michael Persinger in the 1980s; Persinger argued that electromagnetic frequencies might very well facilitate clairvoyance and spirit communication. By combining the Ganzfeld experiment with the Shakti Helmet, Ryan hopes to attract the demons to himself while giving Lara a rest from possession and, through this, a glimpse at a future free from Satan's grip.... Ryan narrates the demons' voices and actions, thereby using "science" to validate their existence for viewers who already believe in them.... Through masterful editing, this stew of religious signifiers, scientific "data," bits of history and testimony, as well as the reactions of the participants are woven together into a seemingly cohesive narrative imbued with meaning for believers.

For nonbelievers, she continues, this narrative is "completely disjointed and utterly absurd" (Walter 2013). A similar situation manifests whenever the onscreen investigators attribute every anomaly of sight, sound, smell, or feel to the subject entity: a bad smell is connected to Bigfoot, a coldness is said to be a ghost. The factualness is manufactured for television. The editing amplifies the reactions and removes any doubt that might manifest. ARIGs that are experienced watchers of these shows will now equate these happenings with their beliefs. If they believe a house is haunted or a Bigfoot is nearby, this is the narrative they adopt. However, after trying field investigations for themselves, ARIGs will admit it's not at all like you see on TV. The hours are long, the conditions sometimes downright awful, maybe dangerous, and the work often tedious, boring or difficult. Yet, guided by what they see described as "evidence" on TV, they obtain results, if not quite as dramatically, that they use to inform the public.

All these witness accounts used to justify media stories and TV shows suggest to some observers that "there must be something to it." But not many dig into the details to see if the core claims and evidence have merit, we just accept them into our respective worldviews. Thanks to the proliferation of these stories, there appears to be an awful lot of "smoke" that leads people to conclude "fire." Instead, a cultural interpretation makes a lot more sense. We've become familiar with ghosts, UFOs, and cryptid sightings through media and social exchanges. The truth is not a single entity or monster, but a complex interaction of perception, reaction, and cultural ideas that keeps changing and spreading through society (Loxton & Prothero 2013).

Sciencey-Sounding Language

The most common means to create a scientific impression by ARIGs was the liberal use of the words *science*, *scientific*, or science-related jargon and titles. Peppering their language with "scientific" this or that––methods, approaches, equipment, procedures, or solutions—is their avenue to add an air of seriousness and legitimacy to their work. Several sites had specific sections pertaining to the "science" of their activities. Commonly used terms (or variations) used in the text include sciencey-sounding words like "frequency," "resonance," "energy," "quantum," "magnetic," "environmental," and "electric." They will preface use of these words by saying, "Experts theorize" or "A popular hypothesis is…" The words "theory" and "hypothesis" are science-speak and are used to convey a more sophisticated meaning to what another observer might call a supposition, a guess, or an idea. Using these academic terms suggests that the idea is supported by other persons. (See the section on *Paranormal Theories*.)

"Quantum" is possibly the most popular sciencey-sounding word used by pseudoscientists in the 21st century. The use of "quantum" by non-scientists should set off alarm bells that the speaker is deliberately attempting to manufacture credibility he does not have. Quantum mechanics, a complex theory that requires immense background knowledge to understand, is so peculiar that scientists outside the field of theoretical physics have a difficult time grasping the concepts let alone being able to adequately explain it. The complexity and uncertain understanding by all except specialized scholars opens a wide space for creative speculation and misuse of the term.

Some scientifically outrageous ideas may sound completely plausible simply because we don't have adequate knowledge to judge them. "Quantum entanglement" is not only purposed as a vague, "hand-waving" explanation for psychic ability, afterlife communication, remote viewing, and extraterrestrial or extra-dimensional visitors, but also invoked heavily for New Age ideas of alternative health treatments. Quantum mysticism is applied to a nauseating degree by celebrity gurus like Deepak Chopra. "Quantum flapdoodle," a phrase coined by physicist Murray Gell-Mann to encompass the misuse and misappropriation of a genuine scientific theory, is a subset of the attempts at sounding sciencey. ARIGs that cite quantum-related explanations should raise immediate concerns that they are faking scientific credibility.

"Energy" is a word often inserted when "thing" or "stuff" would suffice. That is, it is used generically without a specific meaning because it sounds like an inherently meaningful word. (Example: "It has a lot to do with energy."[4]) Spirit or psychic "energy" is supposedly indicated from the responses from ghost-seeking devices. This "energy" derives from the disembodied entity or from consciousness (of a living or a non-living person). However, this is an incorrect interpretation of the scientific definition of energy which is a property of a material object—something made of matter. Ghosts do not appear to be made of matter in the conventional sense and are non-material. A common pleading from paranormalists is that the "energy" of life (referring to consciousness) must go somewhere upon death. A dead body does have energy but all potential is released into the environment when the body is destroyed by death and decay. If the soul or consciousness is something ethereal and not generated by the body and mind, paranormalists contend that this consciousness could remain intact after death. Sometimes, they assert, it gets transmitted to the environment (see "stone tape" or "water tape" theories in the *Paranormal Theories* chapter) to be detected later. Or, the "energy" remains in a place to manifest as an apparition or paranormal activity—a supposition that goes against physical laws that we have already established to be true. Scientific consensus is firm that there is no solid evidence for survival of consciousness after death. No special form of "energy" that survives death. The lack of support for paranormal activity as any kind of energy does not seem to diminish the misuse of terms like "spectral energy" or "ghost energy"—vague "fields" that can be detected by gadgets or certain witnesses. Mysterious "energy" sources are invoked to explain the extraordinary power, effects, and propulsion of extraterrestrial craft. Some cryptid accounts include descriptions of the creature emitting a unique "energy" or detectably energizing or electrifying the environment. Thus, the term "energy" is misused by all ARIGs to explain effects supposedly produced by ghosts, UFOs, Bigfoot and psychic phenomena.

The ubiquitous term "energy" and the trendy term "quantum" are guaranteed to be used in discussion of all these topics but in a scientifically meaningless or absurd way. One group described their effort as "focusing on understanding the underlying environmental or quantum variables."[5] Another described their method as "a parascientific approach to quantum evolution."[6] "A Proposed Scientific Framework for Paranormal Activity" included this explanation of a ghost encounter: "the electric field of a living human may resonate with the quantum state of the solispirit, an intelligent interaction could occur."[7] That last one deserves a prize for being meaningless gobbledygook. The "quantum state of the solispirit" is a made-up term meant to sound both spiritual and sciencey. If it was a real concept, we would expect published research on it and testing of this "resonance." There isn't. The sciencey-sounding language is vague and confusing on purpose with the intent of appearing showy and serious.

Another linguistic habit of ARIGs is their regular assertion of certainty through use of words such as "prove," "rule out," "verify," "undeniable," or "irrefutable" (in terms of evidence). Scientists who publish in journals or write about their own research would never use such loaded words because nothing in science is *ever* certain. A conclusion is considered probable based on the evidence, and only after the researchers calculate the probability of the effects being attributable to chance. Scientific conclusions are statistical, not absolute. Declared certainty—in conclusions and pronouncements—is an indication that the speaker is someone without scientific training, but probably not scientifically illiterate, who is applying concepts incorrectly.

While ARIGs rarely cite current research by the few scientists doing work on fringe topics, name dropping of historically famous scientists occurs frequently. The work of Einstein and Edison are explicitly connected to current ideas relating to the paranormal because technology once promised great strides. Einstein's "spooky action at a distance" in reference to quantum physics is a concept particularly overused and applied to inappropriate situations. Benjamin Radford conducted a Google search to gage the perceived relationship between ghosts and Einstein. It returned almost eight million results indicating that this is a common association.[8] When ARIGs use the "spooky action" concept, the scientific description, along with explanation or citations, is given short shrift because thoroughly comprehending the complexities is beyond the resources of the average person.

Ideas of Heisenberg and Bohr suggested we were on the cusp of revolutionary understanding about causality and materialism (Lyons 2009). ARIGs and others name-drop to give a sense of legitimacy to their ideas even though the work of those scientists and their scientific theories do not connect to paranormal research to any extent.

Many ARIG leaders and independent paranormal researchers who are not scientists consider themselves science enthusiasts. Vince Wilson produced two books—*Ghost Science* (2008) and *Ultimate Ghost Tech* (2012)—even though he is not a scientist. I reviewed these books to determine the legitimacy of the science concepts referenced in them. An ARIG member could reasonably assume that such volumes would represent the state of "ghost science." These books were excellent examples of sounding sciencey. The content was missing references to large swaths of pertinent knowledge about psychology, physics, scientific methodology, and history related to perception of ghosts. Scientific terms were repeatedly misused, and science fiction was regularly conflated with science facts. The recommended reading list contained references that repeated unverified speculative claims and included pop science sources like *The Handy Science Answer Book*, indicating that this was not a well-researched book from a credible expert in ghost science. (For more details, see the Appendix.)

Attempts to assume a role of expert when clearly not qualified to do so reveals a desire to attain a prestigious role in the paranormal community and to reinforce a certain belief system. Wilson had called himself a "parapsychologist" in the past even though his education was through a non-accredited facility. Fortunately, personal communication with Mr. Wilson suggested that he is willing to change his opinions, which is commendable. But these books and ideas are still out there in the public sphere doing a disservice to paranormal investigators and the esteemed concept of science. Four years after *Ghost Science*, Wilson (2012) admits he thinks that most ghost hunting activity is a farce:

"Ghost hunting" has devolved into a social networking hobby and is far from a field of study. The desire for TV appearances, book deals and ego stimulation has exceeded individual inclinations toward any sort of professional commitment. There is no peer review, no sharing of information and no true scientific research being done. It wouldn't be so bad truthfully and mostly harmless if not for the fact that these people with little more training than what they have seen on TV and they are going to people's homes! Can you imagine if this was okay in any field? If your house was being robbed, would you call someone who has seen five seasons of COPS or would you call an actual police officer?

Though I would quibble with several points in his presentation, there is a kernel of truth there. Allowing paranormal investigators with only a mock-up of "training" to be in your home or business that contains valuables (and possibly children), is not a wise decision.

We Use "Scientific Methods"

As I noted previously, around half of groups I surveyed stated they are *scientific* in their methods or thinking. Promotion of this effort is employed most often by stating: "We investigate using scientific methods." ARIGs consider recording and logging environmental changes and anomalies to be collecting data by scientific methods. And they assume that employing objective means such as use of equipment is part of the overarching "scientific method."

Groups that reference *the* scientific method frequently do so in a generic way, for example, describing it as a "procedure for the systematic collection of data through observation and experiment." Methodical, organized, and logical are characteristic of scientific processes, but it's not the whole of science. Being systematic is another way of saying being orderly and careful which applies to activities other than scientific processes.

Several ARIGS stated that they are using "a proven scientific process," "quantifiable and qualitative techniques," and "reliable, scientific protocols." What should reasonably be expected to follow are details describing what those processes, techniques and protocols are, but instead the rhetoric is used more as an advertising ploy to create a sense of seriousness.

Several groups touted their methods as being the most comprehensive or accurate in the field, which is puzzling because there are no standards for comprehensiveness or accuracy in this area of research. Empirical methods were emphasized—that is, collecting demonstrable evidence mainly through equipment or recording audio or visual data, which is not quite right. Equipment use, as described previously, is perceived as scientific and objective because it is viewed as superior to eyewitness observation, feelings, or intuition. I recorded website statements like these examples:

> [We] use equipment that will catch a remarkable display of spiritual evidence.[9]
> Empirical evidence strongly suggests something of a paranormal nature exists in our world.[10]
> Our goal is to disprove until we find empirical evidence to the contrary.[11]

Ghost/ARIGs outwardly state that recording temperature changes, for example, is a *scientific* way of detecting the presence of ghosts, assuming that the equipment is capturing displays of potential spiritual evidence. ARIG members appear to confuse "empirical" with objective, equipment with scientific tools, and gadgets with precise instruments. The concept of being scientific is inextricably linked to the idea of instruments delivering convincing unassailable objective measurements as in Galileo showing the movement of celestial objects through his telescope (Pigliucci 2010).

Many of the statements on ARIG web sites referencing the scientific method, methods, or equipment are *non-sequiturs* or fundamentally fallacious: "[Our goal is] to teach young girls how to use all the scientific methods using various electronic devices."[12]

Using complicated equipment as a way of representing science is similar to using complicated math equations to produce an astrological chart. It seems complex and superficially impressive but it has no logic. One can make a display of food from fake and inedible ingredients look appetizing. If the core reality is ignored, the imitation you are given is ultimately unsatisfying and not useful in the future.

ARIGs can easily fall into a trap of saying the equipment is acting this or that way because of a preferred entity or cause instead of using the data to objectively inform a conclusion. Scientific methods of data collection can tell us "what" is happening but do not tell us why or how. The "why and how" involves a more complicated and difficult process of fitting the findings into the larger display of a sound, defendable explanation. Interpretation of the data involves deciding if it does or does not confirm a hypothesis and how that fits into the theory under investigation.

A few groups have rejected equipment because it preoccupies the minds of investigators or acts as a crutch between the person and the experience. There is certainly merit to this as some investigators become so engrossed with their gadgets that they entirely fail to think about the problem to be solved on a practical level, and may miss obvious solutions to the paranormal problem such as hoaxing by the eyewitness or client.

Some groups also refrain from explicitly using the term "scientific methodology" because they recognize that their techniques and subjects of study are not amenable to scientific rigor. How can an investigator achieve control of all variables in the environment of a family home, a woodland area, farmers' fields, or even a remote cabin where people claim they experience the anomalous? The setting ARIGs work in is far removed from that of a sterile laboratory or a defined, restricted area of observation.

A few ARIGs opt for purely spiritual or occult methods, such as use of psychics. Most often, ARIGs employ a mixture of both objective and subjective approaches. The gadget-heavy, scientific approach is supplemented by spiritual, subjective, speculative, or creative approaches. For ghost hunters, this entails involvement of sensitives, intuitives, or psychic mediums to validate strange equipment readings (and vice versa). Every possible tool, human and mechanical, is used to cover all potential reactions and measurements. This type of approach can be impressive to the client. By employing an approach of both objective science and subjective feelings it appears to the observers that investigators are open-minded, flexible and "covering all bases" with regards to the client's preferences as well.

The group that developed and used the "Ghost Lab" data logging equipment provided a colorful example of pairing objective methods with subjective means. Here is how they describe one incident using tech gadgets: "It was interesting to see how spirits deal with this modern technology. The fact that the entity disliked modern technology during this investigation was confirmed by other psychics on the team."[13]

Because ARIG results are almost never documented in journal articles for others to review, we can't know if experiments were repeated and what those results were. How do these researchers justify the astounding conclusion that the "entity disliked modern technology" without repeated successful attempts to reproduce the results? Remember, science is a collective enterprise, not one person or group's pronouncement that sounds interesting or plausible. But this pronouncement of conclusions without an adequate basis can be found in a large percentage of paranormal observations.

Approach

A mixture of objective and subjective approaches is effective in maximizing results (locating anomalies) and possibly to in enhancing the reputation with the client (or each other). Ghost/ARIGs in this survey nearly universally followed the typical TV ghost hunter method of using gadgets, staking out the location at night, relying on anecdotes and interviews, provoking with objects, words, or sounds, and interpreting anomalies as "evidence" prior to analysis. Testing and attempted re-creation of the reported event by conventional explanations was uncommon (though universally used by skeptical investigators). Use of psychics and sensitives, sometimes even using animals as sensitives, was common. Dowsing, pendulums, protective devices (charms), and prayers were also common. For cryptid and UFO reports, eyewitness accounts were most crucial. The locations were visited and observations were made. Any previous strange story with any remote connection to the area or incident was considered support for a paranormal "hot spot."

The mélange of methods appears to be haphazard with no reasoning or support behind their use. Descriptions of how and why ARIGs gather data are vague and confusing, revealing a fundamental misunderstanding of scientific research:

> Ours is an organization dedicated to the applied science of ghost investigation and supernatural research using a combination of high-tech, psychosocial and spiritual approaches.[14]
>
> We may use "sensitives" to assist investigation towards a scientific conclusion.[15]
>
> In conjunction with scientific instruments, investigators also use natural, clairsentient abilities to study the nature of paraphysical reactions humans experience while being exposed to potentially supernatural phenomena.[16]
>
> [Our group] uses a mix of modern equipment, elements of scientific methodology, psychic ability, quantum theory, meta and quantum physics.[17]
>
> [We] use tools of science and well as our feelings.[18]

Groups will state their intent to "prove" the supernatural via objective means: "We will use scientific means to try to prove that there is a world beyond this life."[19] They aspire to provide "scientific evidence" of life after death.[20] There is an epidemic of inconsistent reasoning.

Misunderstanding of objectivity was also commonly exhibited:

> Everything we do is through a very scientific approach.... We should be using ourselves as the first tool, then, technology ... our minds tell us what's real and what isn't.[21]
>
> [We use a] double blind study method [where] only the lead investigator is aware of the activity history to avoid researcher bias.[22]
>
> We believe in ghosts so you can believe in us.[23]

To set themselves above other ARIGs, some bold groups will declare that while the typical group has low-quality investigation procedures, they are working to raise the standards in the field (of paranormal investigation). Or, they hint that they consider themselves the trailblazers in an evolving field.

> Lets [sic] think of it as a science just being born. With further work by paranormal investigators our research will be eventually accepted.[24]
>
> Science in general looks at the paranormal field still as a "new" or undiscovered science.[25]
>
> [We are] true members of the scientific community rather than hobbyists.[26]

Such statements contradict the long history of paranormal study. Certain ARIGs are convinced that their methods are superior, they will be the ones to demonstrate a groundbreaking discovery to the world. This suggests that they pay attention to what other groups are doing and are actively trying to gain a competitive edge. There are many anecdotes about the competitiveness between groups but little has been accurately documented. Various groups will try to stand out from the others with gimmicks or image, possibly by looking more "scientific" or serious than another one nearby.

Case Documentation

Part of any scientific research effort is reporting your results in a coherent manner that allows others to assess your methods, findings, and conclusions and to put these into context. Access to records and documentation of ARIG investigations are seriously limited. Many groups include reports of investigations on their websites but often they remain private. Group members may intend to write up their findings but, because it is not exciting work, cases often go undocumented. Content and quality of reports are highly variable. Some are very brief summaries or an overview of the group's opinions about the case. Others are detailed including specific dates, times, eyewitness descriptions, environmental and weather conditions, geomagnetic conditions, moon phase, persons attending, specific sensory observations, comments on instrument behavior, whatever measurements were collected, and conclusions drawn. However, most of the content in these reports, particularly for ghost and paranormal ARIGs, consists of descriptions of subjective observations obtained during the investigation. The participants will document their feelings—a touch, breeze, push or "presence," their hair stood on end, they became breathless, cold, nauseated, or sad. They report seeing ephemeral shapes or smelling scents. Any environmental stimulus or perception is deemed an anomaly and related to the paranormal activity. The propensity to assume such entities are out there to discover biases investigators towards paranormal explanations and away from fully considering mundane ones (Pigliucci 2010). Typical reports are of poor quality and not detailed or thorough enough to be useful to someone interested in the case outside the group. Incorrect or imprecise terminology is used. The records of data are lax or incomplete and reflect poor research methodology. The reports do not reach the level of completeness of a scientific paper acceptable for publication. However, that level of quality is too high a bar to expect from enthusiasts who almost certainly have regular jobs and life responsibilities as well as not having any prior training in scientific writing. Writing and formatting such reports is tedious and time-consuming, but necessary. A decent comparison can be made to police

9. ARIG Portrayal of Science to the Public

detective reports that contain enough suitable detail, and providing drawings and descriptions, that would be helpful for another investigator who could take up the case and build upon it. However, we don't see even that level of reporting from most ARIGs.

For a group that strives to be scientific, research must include familiarity with what has already been established as part of the subject area to correctly frame the investigation process and to write well-supported conclusions. This exercise of establishing a foundation for an investigation was not displayed to any significant degree by ARIGs. Citations to any prior documentation or previous research in reports are extremely rare. They frequently do contain reference to the legends about the site (without citation) including words such as "...is said to be," "It is believed...," and "Legend claims..." all of which are red flags to indicate the information is dubious and likely false or incomplete.

Some of the most useful investigations into paranormal claims are those that carefully examined the origin of whatever mystery they wished to shed light upon—a haunting, a monster sighting, a location with reported strange events, etc. Historical searches of primary sources almost universally reveal that the local legend that everyone assumed is true is exaggerated, changed, or without basis in fact. The story became lore and was passed on without verification. That the story is old and well-known is used as a surrogate for responsible fact-checking: everyone knows this story, there must be something to it. From typical examples of ARIG reports that are available online, I failed to find one report that would be independently judged to be thoroughly researched to the extent that the investigators examined the origin of the claim and established references for what is known. Such work requires long hours searching newspapers and documents, locating previous witnesses, or chasing down details and verification. If done well, such writeups are useful and deserve publication in a place more prestigious than a personal web page.

ARIGS will visit certain places as "meccas" of known paranormal activity. Dozens, maybe hundreds of groups, have independently investigated locations like the Gettysburg battlefield, Myrtles Plantation, and infamous "haunted" places. Popular Bigfoot areas include northern California locales like Willow Creek. UFO seekers flock to the eastern Nevada desert for sighting flying objects. The resulting logs or reports from these visits, if any are completed, are not detailed enough or openly shared with others to be of any use in comparison. The collection and organization of independently gathered data from these well-known locations is not curated.

UFO and cryptozoological groups have databases of reported eyewitness sightings which is a promising concept. The records in these databases typically include environmental conditions, location information, eyewitness descriptions, and occasional drawings or photographs. The intent of the databases is to use the records to plot perceived movements and trends in sightings. Databases of UFO sightings through MUFON and Bigfoot sightings through the Bigfoot Field Research Organization are often self-reported, not checked for accuracy or veracity, and are incomplete. The ARPAST database was the only such collection of records for ghost research I located in the U.S. It is not readily evident that any of these databases are used in scientific research projects, although there are some instances where they have been useful.[27] I found no serious effort to use or assess this data in a statistical way that has been documented by amateurs. The exception is for UFOs which were statistically assessed by the USAF in the studies from 1952 to 1969. Other ARIGs may consult online databases for information on where to investigate next based on reports that include location details.

Paranormal Theories

The use of the word "theory" and descriptions of ideas that are called "theories" are common features of ARIG discourse, and paranormal-themed discourse in general. A *theory*, as used within scientific contexts, is different from common non-scientific usage. This important distinction is not embraced by ARIGs. Instead, the word is used as an important credibility signal. Presentation of a *theory* sounds impressively scientific even when what is described is a speculative guess.

A scientific *theory* is *a well-confirmed explanation of nature*. It is different from a *hypothesis*, which is an idea that has been derived from observations but requires testing to be confirmed or rejected. When faced with uncertain explanations and especially difficult or complicated problems to solve, investigators must hold multiple working hypotheses and be allowed to maneuver between them to find the best fit for the evidence. A theory is more than a glorified hypothesis, it's the entire pie rather than just one slice, and has been well-tested and passes the tests. The theory serves as a model to explain and predict nature (Dewitt 2004). The most popular examples of scientific theories include evolution by natural selection, relativity, gravity, plate tectonics, molecular chemistry, and the germ theory of disease. These models explain various observations in terms of the larger framework of nature. A well-supported theory allows us to explain nature as it all fits together into a coherent picture.

Gigerenzer (2009) describes what he calls "surrogates for theory" with regards to psychology. He says that psychology seems split into two camps: one, where a well-formed theory acts as a model that can be used to test predictions with experiments, and the second camp that runs experiments with no meaningful theory but a crutch of an idea that serves as a surrogate for theory. With the latter, almost anything passes for "theory"; some are more accurately called "hypotheses" but even this word strongly suggests potential support for the idea. In amateur investigation topics like hauntings, cryptozoology and ufology, creative explanations are rampant. The explanations may be conspiratorial, supernatural, or otherworldly. They do not fit the category of "well-formed" at all. And, they certainly do not rise to the level of scientifically tested frameworks for explanations.

There are currently no acceptable, supported *theories* to explain the concepts of ghosts, UFO sightings, or cryptids, individually or collectively. Instead, the explanations for these reported experiences are multiple and various. One explanation does not suffice for all reports of ghosts, for example. To say that there is an overarching "theory of ghosts" is nonsense. The following is a cursory look at the most common suggestions for explanations proposed by ARIGs and those interested in these phenomena. The lists are not inclusive as there exists many other creative ideas that are referred to as "theories," but these provide a flavor for the preferred explanations by ARIGs. As you review them, consider the evidence we have for each one and how that does (or doesn't) correspond to our established, well-tested framework for how nature works.

Ghost Theories

Spirits of the dead. This is the most common interpretation—that the energy of a living person can somehow manifest itself after physical death. This concept has been heavily reinforced throughout human history and currently is an extremely common motif in popular culture. The idea that the "soul" or consciousness can separate from the physical body is so ubiquitous that many people just accept it as true.

Demons and angels. Supernatural entities visiting humans are stories of legend and ancient myth. But in modern times, many people believe that demons or the devil himself can manifest to a person or possess a human or animal. Angels are popular especially to explain cases of recovery from illness or protection from danger. Prior to the scientific age, ghosts were thought to be demonic tricks on the earthbound or an angel delivering a message.

Recordings played back in the environment. ARIGs often refer to the *stone tape* theory or *water tape* theory. If they subscribe to this idea, which sounds sciencey, they will associate the bedrock, building material, or soil of the haunted location with a certain type of rock or mineral, most commonly quartz or limestone, concluding that the rock can "record" or capture an emotional imprint like a tape recording. Flowing underground water or surface streams also are said to retain a memory of an event that can be played back to receptive people in the right circumstances. These ideas can be traced to the concept of psychometry—that a psychic impression is recorded into an object or substance (Maher 2015).

Quantum entanglement. The non-intuitive weirdness of quantum theory has been used by some to show that ghosts are a product of physics we don't fully understand. This concept allows ghosts to be the disembodiment of a living person (or animal), a somewhat more complex version of recordings of emotional "energy" in the environment—it's working at the subatomic level, where things get very strange.

Projection of the mind. Some people claim to see ghosts while a person next to them does not see the same. One suggestion is that the mind projects the image of a ghost and only those people who have a special ability for psychic communication, a "sixth sense," will receive the image. This explanation for ghosts used to be more popular in the early days of research. A related concept is that of thought-forms or *tulpas* where one or more people can create an entity via their own mind power. Medication, drugs, illnesses, or mental conditions may cause hallucinations that appear to be very real.

Inter-dimensional visitors. Ghosts may be interpreted as beings from another dimension or an alternate universe who have managed to traverse the boundaries that we humans have not yet figured out how to cross.

Time slips. There are stories from witnesses who report they appeared to have gone back in time, where everything and everyone around them is not as it should be. Some put forth, often referencing Einstein, that time leakage can occur in certain places. The ghosts are temporary manifestations of the time leakage associated with that unique location serving as a portal between ages.

Environmental conditions. Mists and clouds may be perceived as ghosts by those inclined to believe in them. Some research has focused on magnetic fields, solar wind, ultrasound, and geologic stress to assert that environmental conditions can cause people to experience a real or perceived ghost.

Misidentification. Witnesses can conclude "ghost" by misinterpreting real people, smoke, reflections, light anomalies, shadows, etc. *Pareidolia* is the tendency for humans to find a pattern where there may not be one. Therefore, a mist or shadow can be interpreted as a human form.

Technical anomalies. These are misidentifications that have become extremely common as technology use has become ubiquitous. Most viral "ghost" stories today are the result of camera glitches, lens flare, and video artifacts. Double or slow exposures have been perceived as ghostly evidence since cameras were invented. ARIGs interpret tech glitches as spirit communication or paranormal effects.

Hoaxes. Always an option that must be considered. Hoaxes have occurred throughout human existence and will continue to be carried out. Today, ghost hoaxes can be created by anyone with a smart phone app and a few clicks.

An example from one of the dozens of phone apps that allow the user to place a "ghost" into any image on their phone. The app used for this image was Ghost in Photograph by Leeway Infotech, LLC, and is free to use. Photograph by Kenny Biddle.

UFO Theories

Extraterrestrials. The most pervasive cultural idea about UFOs is that they are alien craft and that the pilots are from other planets or galaxies. These advanced civilizations are visiting earth to study, use, or warn earth inhabitants. This concept—referred to as ETH or the "extraterrestrial hypothesis"—has been a basis of science fiction for decades and has become ingrained in American culture to the point that a large percentage of the population thinks the government knows about ETs but isn't telling the rest of us.

Ultraterrestrials. UFOs may be vehicles of civilizations hidden here on earth—inside the earth, underground, or undersea.

Time travelers. UFOs may be advanced humans from the future visiting modern times. Support for this idea is said to be that the typical "gray" alien looks to be a highly-evolved human form.

Inter-dimensional visitors. As with ghosts, above, UFOs can be the transportation between other dimensions or alternative universes.

Living things in the sky. UFOs are not mechanical craft but organisms. This idea is currently gaining in popularity. Examples of the organisms include "rods" (which are artifacts of digital cameras resulting from flying insects or birds) that some people also call "sky fish." Also, appearing more frequently are "jellyfish" UFOs that some people think are biological entities. Images of such creatures are manufactured hoaxes.

Natural phenomena. UFOs are typically described as lights. It is well known that planets, the moon, and bright stars have been misidentified as UFOs. Electrical or atmospheric phenomena (including several kinds of lightning, electrical discharges, ionization effects, cloud formations, and optical effects in the sky) can be strange enough to be labeled UFOs. Meteors may also be interpreted as UFOs crashing to earth. Pilots can even be fooled by light reflections while flying that appear to be solid objects following them. Even flocks of birds can reflect light in a strange way and appear odd to a viewer.

Man-made technology. Along with natural sky phenomena, there are a vast number of man-made things in the sky that people may not be able to readily identify. Top secret technology must be tested, usually in remote areas of the country, and we don't know what's being tried out. Weather balloons and scientific equipment can look strange, reflect light, and appear to be intelligently directed or going against the wind. The most obvious things in the sky are typical aircraft. Now remote-controlled drones or quadcopters have unique light displays. Satellites, rocket launches, and the International Space Station can be surprising to witnesses unaware of how these appear in the night sky. There are also floating lanterns, Mylar party balloons, and plastic bags that float incredible distances, changing shape, resulting in UFO reports.

Projection of the mind. UFOs have historically been associated with military activity and wartime. One theory is that lights in the sky may be a "thoughtform" created by the anxiety of a citizenry prepared for war. Or, UFOs are collective delusions by people afraid of technology or fearful of environmental destruction.

Technical anomalies. Again, as with ghost theories, camera glitches, lens flare, and video artifacts may be interpreted as unidentified flying objects. Long exposures turn flying insects into high-speed alien craft.

Hoaxes. Always an option that must be considered. Hoaxes have occurred throughout human existence and will continue to be carried out. UFOs can be added to any photo via smart phone apps. Impressive computer animation of UFOs are all over video sharing Internet sites.

Cryptid Theories

Real animals to be discovered. The basis of cryptozoology is that there are genuine new species to be recognized that scientists have not officially catalogued on the land, in water, and in the air. This includes animals or other species of hominins thought to be extinct but still surviving in certain areas. And, it includes prehistoric survivors like plesiosaurs, pterosaurs, dinosaurs, or Pleistocene-age fauna. There is also the belief that specific types of animals, such as large cats like mountain lions, live in areas where they are not expected to be. In some cases, cryptids may be exotic animals released or escaped into the wild. This theory of unknown animals out there to find is reinforced by regular discoveries of new species. However, the new discoveries never approach the level of interest as would a Bigfoot or sea serpent. Carcasses washed up on shore or found by accident are commonly called "monsters" and people assume they are new species as they defy immediate explanation. They are frequently identified later by experts as common species.

Spirits or entities. Many ARIGs connect cryptids to myths and legends of native people. The animals may not be flesh and blood physical beasts but gods or demons in animal form, guardian spirits, or protectors.

Zooform phenomena. These are materializations of animal-like creatures. They are said to be conjured by human thought, also known as *tulpas*. Because they are not biological entities, they have supernatural behaviors.

Aliens or alien pets. Some cryptids are so weird in their descriptions that people think they are not from this world. They are aliens sent to explore earth, or, they are alien "pets" left here when the UFOs exited. They can also have paranormal qualities such as instantaneously blipping out of existence or associated with a UFO sighting, their assumed transport.

Mutants and hybrid animals. Animals found long-dead or reported from fleeting observations may not be readily correlated to their familiar living state. Some illnesses, injuries, genetic abnormalities or anomalies and decomposition effects render a common animal to looking abnormal or very weird, preventing immediate identification. When witnesses are not familiar with these conditions, they assume that's what it is supposed to look like. Trying to make sense of an identification based on what they know, amateurs resort to concluding the creatures are some genetic mutant or a hybrid monster. Media reports will latch on to extraordinary claims that the weird sightings or remains are from experiments gone wrong or escapees from a secret government testing lab.

Misidentification of real animals, people or things. Many tree stumps have been labeled "Bigfoot" in photographs. Any dark object among the vegetation that vaguely resembles an ape-like form becomes a "blobsquatch." Any decomposing carcass or hairless carnivore is labeled "Chupacabra." Waves, birds, floating debris, and swimming deer are reported to be lake monsters. Pareidolia is rampant in cryptozoology. Night-vision cameras can capture real animals and people with enough vagueness to be counted as unusual. Unfortunately, many witnesses and even researchers put little effort into identifying the creature as something already known, and quickly resort to making it mysterious.

Inter-dimensional visitors. As with UFOs and ghosts, the "visitors from other dimensions" is a catch-all, open-ended, no-rules explanation. A portal to the other dimension allows the animal to disappear without a trace, as is often reported with cryptids.

Hoaxes. Hoaxes are part and parcel of cryptozoology. From people modifying real animals to look odd, releasing non-native animals to freak out the locals, fake footprints, photos and videos including people dressed up in costumes, the hoax must always be considered. Hoaxes make the news and give the hoaxer a jolt of satisfaction.

From all these examples we can see that most paranormal theories are speculative, vague, and esoteric ideas that are highly creative and interesting but are difficult or impossible to test and have no prior plausibility to be true. Speculation is useful to create new ideas, but the follow-through must occur whereby the idea blossoms or dies. Novel ideas have come out of science but the scrutiny of such ideas takes a huge effort (Dolby 1975). ARIGs' creative theories come with ample imaginative thought and are delivered with sincerity and commitment, but little to no evidential support. Instead of saying "I don't know," ARIGs create an explanatory framework that is satisfying to them and an audience may accept it. Many of these theories sound sciencey and plausible to those who don't know much about physics and biology. Yet, the evidence used to promote such fringe ideas is not convincing to professionals in related fields, not helped by their inherent implausibility, they are completely rejected by mainstream science.

Paranormal theories meant to explain weird experiences ultimately create more problems than they solve since we would have to rewrite much of what we already have established to be true. The solution ARIGs use to maintain their favorite theories is to skip the scientific gauntlet and appeal directly to the public. When these colorful and interesting (but baseless) ideas are picked up by the public, they gain popularity. Reliable knowledge and scientific conclusions are considered boring, tossed out for the more exciting and controversial "theories" which can be more aptly called "wild guesses."

Contacting Scientific-Minded Groups

To provide greater resolution into the reasoning why certain groups exhibited strong scientificity, I contacted via email thirty groups that clearly advocated a "scientific" method in their name and/or website presentation to request additional information about how they apply "science" or consider themselves "scientific." Of the thirty, eight replied to the questions listed below.

- What is it about your methods/procedures do you consider "scientific" in nature?
- Do you utilize methods that are non-scientific? Please describe what these methods are.
- Are any of your members trained in scientific methodology? That is, do they have experience in conducting scientific research outside of paranormal investigation?
- What forms of data could you supply to the scientific community to consider?

Of those who did reply, they exhibited a guarded attitude and suspicion in speaking to me. When asked directly about the scientificity of their groups, the representatives included qualifying information or, surprisingly, immediately retreated from a strictly scientific methodology: "I wouldn't say that are [sic] methods are necessarily scientific," one wrote to me, even though that was clearly stated on their website. One group, who stated that they were unbiased but that use a "scientific approach with a religious basis," responded to a question regarding what it is about their methods/procedures that is considered "scientific" by replying, "Some of our scientific methods are trying to find an explanation for what may have occurred by in depth [sic] research and investigation to try and explain and/or re-create what may have occurred under controlled conditions." I did not, however, see any evidence of in-depth research and controlled conditions in their reports. This suggests they were not thorough in their documentation or they regularly did not fulfill those standards.

Another representative qualified their data sets by noting they may be unreliable or mistaken. They understood it may be impossible to attribute any event to paranormal activity. Another noted that their data "must largely be accepted on trust—trust that they haven't forged or altered it." They also admit that their methods are "experimental, untested and unverified." One ARIG leader expressed the feeling that the scientific community would never consider any of the evidence and that there cannot be scientific proof of an afterlife. These responses were odd in that they deviated from their public web statements. How quick they were to back down when called on their scientific credentials! Those who admirably admitted that to be strictly "scientific" is difficult and that their results will likely not convince the scientific community recognized they are not recording evidence of paranormal activity in a consistent, repeatable manner. And no one collection of data has been enough to confirm a haunting to a wide audience. But paranormal questions don't easily lend themselves to direct testing under controlled conditions. To do so is extremely difficult, time consuming, and would cost a great deal to carry out.

A few ARIGs are adamant about being scientific and do not demur, insisting that they have very high standards. One ARIG leader, who is not a scientist, stated that she teaches classes in the scientific method for paranormal investigators. There are no state laws against impersonating a credentialed scientist in this manner but it smacks of pretense.[28]

When asked what evidence they *could* provide to the scientific community, groups acknowledged the shortcomings of their evidence or provided just vague answers. ARIGs do not typically submit evidence to the scientific community. Only a rare few groups have members with any connections to an academic. ARPAST is one well-established group that portrayed a high degree of scientificity on their site. I contacted them as one of the thirty groups but they did not respond to my questions. ARPAST is unique in that they state on their site that they are "collaborating with doctors, scientists, universities, and other legitimate science-based organizations to build and utilize a research database." Access to the database is restricted to "legitimate scientific research organizations only." I requested access to this database under the auspices of this research project by completing the application as required but received no response. No names or credentials regarding the aforementioned professionals noted could be found nor were any citations given to suggest use of the database for research. Thus, I could find no evidence that backed up the claim that they are collaborating with scientists, nor did I find results of such collaboration. If they indeed are doing that, it's either not been useful or results of such research was not shared.

In their replies to my questions about why they consider themselves scientific, ARIGs indirectly revealed that they were fully aware they were doing a good bit of acting for the sake of looking more credible. They understood that this was not formal science, that it was an approximation. ARIG spokespeople appear confident and comfortable appealing to the public's sense of what is scientific (based on television portrayal of science), but they equivocate when confronted by a knowledgeable inquirer. They perceive this weak spot even if they don't admit it to each other or themselves. When probed to think about it, they show awareness that a more accurate scientific process would be more rigorous, far more difficult, and perhaps a lot less fun than what they do now. Ghost hunters already find it boring to listen through hours of audio recordings. Cryptid- and UFO-seekers sit out on cold nights in remote areas. It's not glamorous! They seem to wish to do science the best they can but only to a point; then the effort becomes too laborious or impossible.

10

Methods and Evidence

"Evidence" is anything put forward in support of an argument. A broad term, many different kinds of evidence can be identified depending on the situation you are in including facts, judgments, observations, experimental results or collected objects. There is strong evidence and less-strong evidence. Part of collecting evidence means paying attention to the context in which the evidence is collected. Evidence can be high or low in quality and anywhere in between. *Empirical* evidence, such as experiments and recorded observations, is required in science. But empirical evidence can be worthless if not scrutinized, controlled, and focused. Thus, the method of examining the data is critical. In science, the empirical method means moving from observed facts to explanatory theories. (See the section called *What Is Science* in Chapter 8.)

Not all ARIG websites have evidence from their investigations available for public viewing or even have it in a format available to non-members. Several sites state concern for their client's confidentiality and display no results without permission. Most sites do have one or more categories of evidence for public access, typically photographs, audio recordings, and video clips. For some ARIGs with a strong belief and a lesser emphasis on scientificity, it's not important to ask for evidence or proof of a claim. As with religion, it's socially improper to ask a person to justify a belief (Denzler 2003). Belief is often deemed to outrank any "proof" for many people. Invested believers don't require evidence and often don't care what others think about them.

While viewing the evidence, it's important to distinguish between data and interpretation of that data. Equipment used by ARIGs are tools to measure environmental conditions. The equipment itself does not make a conclusion; it produces data that must be analyzed and interpreted before it can applied to a certain conclusion. Several pitfalls lie in the path from obtaining the data to making conclusions from it, especially if the context of the evidence is ignored. What follows is a brief summary of the various examples that appear on ARIG websites as evidence and the means by which they collect the evidence typically attributed to paranormal causes.

People as Instruments

The typical ARIG process often described on their websites consists of eyewitness interviews, site visit(s) with equipment setup, collection of data in usually one, but possibly multiple, days/nights, analysis of data, presentation of the results to the client, and a write-up or record of the investigation. This process varies depending on the data set collected (e.g., in the case of a UFO sighting or no-data-collected, there will be no presentation of results). Since the focus is on an unusual phenomena, ARIGs are faced with the problems of non-reproducibility and subjective interpretation. In other words, this is a difficult subject to study in the real world that is messy and influenced by countless factors outside the investigators' control. However, throughout time to present day, the evidence most used by ARIGs is the least useful in terms of getting to the truth—eyewitnesses testimony and anecdotes. Personal experience and testimonials comprise almost *all* the persuasive evidence in books on these topics. It is simply recorded and accepted at face value. Testimony is privileged over empirical data by most people (Blancke et al. 2016). They will intuitively trust the witness and not examine the claims of the witness too deeply, if at all. A common instruction given to ARIG investigators by leaders tell them to trust their feelings, that "you are the best instrument." It's true that you are the best instrument to produce subjective feelings, but not objective results. Extensive experiments and large formal and informal studies have, without a doubt, determined that humans are very unreliable observers. As an example of the real-world ramifications of this issue, consider the people who were put in jail by convincing eyewitness testimony alone only to be later released by DNA or other physical evidence that acquitted them. Yet, we've all heard people say "I know what I saw!" The hard truth is that they *do not* know what they saw. They have *interpreted* what they saw. Their senses gathered information and their brain built an impression of it based on the individual's existing worldview and experience. Then, it was stored in memory and later related to another person. Errors can occur in each of these steps to foul the factual account.

Scientific research is designed to minimize mistakes and bias of personal perception as much as possible. People are "complex instruments" of observation because of our mistaken perception and misinterpretation (Morrison 1972). Yet, ARIGs commonly note that they use themselves as the primarily tool, erroneously assuming that humans are a reliable instrument for recording data through perception: "We should be using ourselves as the first tool, then, technology ... our minds tell us what's real and what isn't."[1] This is precisely upside down.

Special weight is provided to those witnesses in professional positions such as policemen, doctors, pilots, or scientists. Even stories told by respected witnesses like pilots, or trained observers like scientists, are prone to be mistaken or embellished to the point of being wrong (Loftus 1996). While some may be trained to observe in their area of expertise, such observation skills do not preclude errors in perception or translate to other types of observations, especially under less-than-ideal circumstances. Even though it is a commonly circulated analogy, human memory is not at all like a tape recorder.

Some people are what Lamont (2007) calls "believers in waiting." When asked, they will say they are skeptical of the existence of ghosts, UFOs or strange creatures but are ready and willing to believe given their own experience. Because the average person is not at all well-versed in understanding how flawed our perceptions and memories can be; if they have an experience that they interpret in a certain way, that interpretation becomes a memory. That memory is re-formed each time it is related. Because humans are cognizant of the context of a situation, if the experience is later related to a sympathetic listener, the presentation will be different than if related to a doubtful listener or a less enthusiastic audience. How the event is subsequently described to others changes the memory. Famous paranormal cases are notoriously inconsistent over years as people's stories change or they "remember" something new. Telling your eyewitness testimony to a listener can be less an objective account of what happened than a personal catharsis. In that sense, testimony loses important information and gains possibly misleading information for the person who is collecting memories of eyewitnesses as evidence. Anyone who conducts any type of investigation should understand the unreliability of eyewitness testimony and memory. Use of individual personal accounts as the main evidence, requires understanding of the limitations and problems that can occur with perception, retaining memories, and retrieving and retelling from memory. An indispensable volume on how this process can go wrong is *Eyewitness Testimony* by Elizabeth F. Loftus (originally published in 1976, now updated).

Disagreement is heated between those who accept these anecdotes at face value and the more skeptical audience who knows that this form of evidence is sometimes useless to determine what factually happened. The reliability of eyewitness testimony and anecdotes forms an unbridgeable chasm between those who accept or do not accept paranormal claims. For some, the anecdotes are all that are required. For others, the stories are unverified and not reliable as evidence at all. The truth may be somewhere in the middle as anecdotes over time might suggest that something interesting is going on but the anecdotes themselves are not sound data. Anyone who says that our minds are arbiters of what is real (or not) is misguided and could be dangerously misled by that assumption into making flawed conclusions about many aspects of life.

Those who are unaware of how we fool ourselves by finding meaning in patterns when there is none are prone to being the first to be fooled (Dewan 2013). The most obvious way we make mistakes is by seeing a face or figure in noise, a common error in amateur paranormal research results. As noted in Chapter 9, "pareidolia" is the name for the human tendency to perceive "vague or obscure stimulus" as something "clear and distinct."[2] This illusion, which we can blame on our brain trying to make sense of things, accounts for ghosts and spirits in photos, Bigfoot among the trees, giant flying objects in the dark sky—all of which are presented as evidence that these strange things are real. It's also referred to as "matrixing" (when we can see a human shape in a vague image or hear a recognizable phrase in garbled sounds such as with EVPs) and "apophenia" (making connections and meaning out of random events or objects). These experiences can be surprising and compelling. When a photo or video feeds a desire to believe in a certain explanation, some people have all the proof they need and no other explanation will ever be accepted. To look deeper and question this type of evidence feels uncomfortable for both parties—the skeptic and the experiencer. It feels insulting because we are questioning their ability to perceive reality. We all mess up—it's human. But we can recognize this and correct for it. The first step is to admit it and understand why it happens to everyone.

Technology

Technology plays a strong role for most ARIGs. Several groups expressed the notion that new technology is the key to making breakthroughs in paranormal research. Most ARIG websites had specific information about and, typically, images of, the equipment used in an investigation. Equipment commonly utilized in ghost investigations are cameras (digital, film, video, night-vision, infrared), electromagnetic field detectors, audio recording equipment (magnetic and digital), temperature gauges, laptop computers and associated software. Additionally, some groups use more expensive specialized equipment retrofitted or designed for attracting or communicating with spirits such as ion generators and white noise devices.

Cryptozoological investigators include FLIR cameras (forward-looking infrared or "heat sensing" devices), parabolic microphones, and game cameras as part of their necessary gear. They may carry kits to collect biological samples or cast prints.

Ghost/ARIGs are reliant on equipment and technology as the key to gaining solid evidence. Harvey explains in Jenzen and Munt (2013: Chapter 2) that these kinds of records from technology are no better than testimonials since interpretation is still based on belief in a particular explanation. The sounds, images, and light blips can tell the truth or a lie or be completely misinterpreted. When the users are not aware of how the technology works, its limits and failures, they will conclude an explainable anomaly is, instead, a ghost.

Equipment is a seen as a necessary accouterment to appearing objective, "scientific," and professional. Equipment that collects empirical data seemingly validates the subjective observations of the investigators. Unfortunately, artifacts of the instrument can be misinterpreted—many paranormal investigators do not understand the technological limitations of environmental meters, cameras, and recorders—and so anomalies that might be explained in terms of the technology (camera glitches, natural variation, ambient noise, etc.) is used as evidence to support a preconceived conclusion.

Visual Evidence

Still and video cameras (mostly digital) are a considered a necessity for field investigations of all types. Visuals are a must for illustrating results. ARIGs make attempts to capture anomalies on "film" and supply these often ambiguous, frequently misinterpreted moments in time as part of their collection of evidence. They will frequently guide you as to what to look for and suggest what you should see. The most vague and ambiguous examples are used to support paranormal claims.

Cameras Lie

Cameras can lie through countless unusual effects, glitches, that result from the mechanics of the device, lighting, settings, and the environment. Photos can lack scale so the viewer can't judge size, distance, or details. Eyewitness photographs of aerial anomalies are particularly prone to scale misinterpretation since photos of the sky often contain no other objects for which to judge against the anomaly and that anomaly is almost always too

10. Methods and Evidence 161

far away to provide decent resolution. There is no way to determine if the object in the frame is far away and large or close and small. Aerial anomalies that might be counted as UFOs are seen and then are gone too quickly for the witness to get any visual evidence.

Bigfoot is the ideal example of the classic bad portrait. It is impossible to find a photograph of Bigfoot these days that isn't blurry (or an obvious fake). Because of the jumbled, variable, wooded setting in which mysterious animals roam, any black blob in a visual recording could be interpreted as a Sasquatch. These amorphous blobs are perceived by some to form the shape of the iconic North American wood ape. Blobsquatches are the extreme example of how, when observing a photograph, recording, or some other evidence, the ARIGs will shoehorn it into their preferred explanation. Many have been revealed to be tree stumps or rocks that are simulacra of hiding creatures.

Some crypto groups have set up remotely triggered cameras left for stretches of time at a location thought to be an active game trail. Their goal is to capture the local animal population, and hopefully, find a cryptid or other surprising animal walking among them. Trail cameras have been successful in capturing evidence of the presence of unexpected and rare species, and a few pranksters, but never a Bigfoot—the ultimate prize. Trail camera images are frequently overexposed by flash, distorted by movement of the animal, or capture the animal partway in the frame. Since they take images at set times or when a subject is detected by movement in range, the best

An example of a typical trail cam photograph taken at night with an automatic infrared flash that detects movement. The subject's shape is distorted and details are blurred due to the slow shutter speed. Photograph by Kenny Biddle.

angles are missed. Vandalism or theft of the equipment frequently occurs. It's also difficult for ARIGs to retrieve cameras installed in very remote locations that may experience severe weather. While game/trail cameras seem like an ideal way to capture evidence, they rarely succeed in doing so. Or, as some cryptid-hunters suspect, Bigfoot knowingly avoids them.

As with other photos, game camera images are not definitive but require an interpretation by the viewer. If the viewer is motivated to accept one interpretation over another, the photo (or video) anomaly will be interpreted to be whatever the viewer wishes it to be. The audience for this media is provided with annotations like circles, arrows, or drawn-in outlines to help lead you to see their interpretation.

ARIGs often work after dark maneuvering with flashlights. Ghost/ARIGs will deliberately turn off all the lights in a building under investigation.[3] This can enhance the appearance and subsequent misinterpretation of several artifacts in the photos that the investigator did not perceive in person. Many photographs claimed to be of spirits or otherwise paranormal are mists, condensation, reflections, shutter effects, shadows, offset duplicate images or obscuring shapes in the frame (including thumbs). No camera anomaly is more associated with ghosts and hauntings than orbs. *Orbs* are semi-opaque ball-like artifacts that commonly appear in digital photos. They

Orbs are artifacts of the camera's flash reflecting off dust particles, precipitation, or insects. This array of orbs was created by walking back-and-forth in a grassy yard and randomly taking flash photographs. Photograph by Kenny Biddle.

are not seen by the naked eye at the time of the photo but are created by digital photography, greatly enhanced by flash. ARIG websites frequently include photographs of "orbs" presented as evidence. The capture of orbs is so common that it remains irresistible to researchers to remark upon a strangely placed sphere that shows up in a photograph. Orbs are known to be caused by reflections of light (especially camera flash) from dust particles, insects, water vapor, or precipitation. However, a subset of pro-paranormal investigators claim that particular orbs are indicative of "spirit energy" present. They provide advice on discerning the difference between a paranormal orb and natural one. Orbs are connected to places where unusual activity is reported. They even appear on video where they move in what is perceived to be *intelligent* ways. Several paranormal ARIGs now avidly disavow orbs as evidence because they have such a demonstrable, prosaic explanation. I came across several websites that posted a "badge"[4] that stated "No Orbs."

Videos

Video clips made by Ghost/ARIGs are also taken in low-light or dark conditions by using night-vision cameras. These cameras give the spooky green cast to the playback that is now readily identified with reality-ghost TV shows. Video clips collected as "evidence" most often show group participants active in some portion of the investigation, and sometimes capturing unusual movement or behavior of equipment, or moving objects, lights or shadows. As with still photos, videos require interpretation.

Cryptozoological evidence may be captured in daylight to document the environmental conditions or anomalies found. Occasionally, a video will show some obviously mobile object or animal far in the distance, often obscured by trees or submerged in water. Crypto/ARIGs also use infrared or night-vision recording devices outdoors. FLIR (heat-sensing) cameras register warm objects and are handy at spotting a hidden animal (or person) at a distance. But in many cases, because of the distance or the surrounding natural environment, lack of clarity makes it impossible to discern what is in the frame.

While still cameras have their notorious orbs, digital videos have their "rods." Rods appear to be animals "swimming" through the sky. A few fringe researchers have concluded that rods are actual cryptids themselves! However, these rainbow waves zipping through the frame are readily explained as normal winged animals, usually insects, whose wing-beats sequences are captured within the frame, distorting the image. In 2012, a local news station in Denver capitalized on a ridiculous story of UFOs caught on film around a field outside the city. Why did people only see this mysterious event through a video camera, only at certain hours, and not while standing there? Because they were small flying insects distorted by the camera.

Hoaxes

Photos and videos of paranormal activity have been hoaxed since the camera was invented. With the availability of low-cost devices and digital software, a visual hoax must be a primary consideration for any so-called evidence. Artifacts are even more probable. We see what we believe. Every photograph and video must be considered a potential hoax. There are thousands of hoaxed creature, ghost, and UFO photographs and videos. In some you can see the badly fitting costume, or you can tell it's a model. For paranormal claims, ghostly mist is only exhaled smoke or manufactured with a spray bottle. Need to move some objects around or suspend them in the air? Clear thread is handy. UFO hoaxes range from DIY fishing line to mimic flight to professional computer graphics. You can create your own community panic with a balloon and some LED lights from a discount store, though I don't recommend you try—it can get you in trouble.

The word "photoshopping" has become a verb to show how widespread it is to manipulate photographs. If learning how to use these complex programs is too much, smartphone apps are fast, cheap and easy ways to fool your neighbors and the local media into thinking you captured a ghost, UFO, or an image of Jesus or Mary. It's simple to take a normal picture and insert any abnormality to your liking. There have been countless cases of television stations and news websites reporting photographic enhancements as "mysterious" or real because they do no investigation themselves and underestimate the human desire to pull a fast one.

Several individuals are known to be serial hoaxers claiming they have proof of UFOs, a perfect image of a ghost, or a Bigfoot body. People will hoax for a variety of reasons—to see how far they can get with it, to garner attention, or to secretly laugh at the gullible. Never underestimate the human capacity to be fooled and the ingenuity of hoaxers who do the fooling.

Audio Evidence

EVP

Audio evidence from ghost/ARIGs is prevalent in the form of EVPs (electronic voice phenomena). EVPs may be recorded via magnetic tape or digital recording devices, computer microphones, or on video recordings. The assumption in the acceptance of EVPs as evidence is that an intelligent, disembodied entity has been able to affect the recording device to communicate, that the sound is actually the voices of the dead. The entity—located in another dimension or world, wherever you go after life ends—it is surmised, may not have the capability to generate a sound within the range of human hearing but can emit "energy" to make word-sounds that are captured on the recording device. While some of the recorded sounds remain unexplained, EVPs have not been shown to be actual voices of the dead. Alternative explanations just as unreasonable include sounds from aliens, trans-dimensional communication from the future, and living persons' psychic messages. Such hypotheses can't be easily tested. EVPs, as envisioned by the originators—Jurgenson and Raudive—were simply mysterious voices. Those voices may be normal-sounding, distorted, whispers, shouts, robotic or alien-like. The history of EVP, or "Raudive voices" as they were first called, is complicated and fascinating and beyond the scope of this book. See Sconce (2000) and Leary and Butler (2015) for more on EVPs. There have been few controlled studies of EVPs to determine if they are anomalous at all.

Today's ghost/ARIGs often are unaware of the origin of EVPs except that they've seen it used by TV ghost hunters. Capturing EVP is accomplished by letting the recording devices run and see what gets caught. Alternatively, it can involve direct questioning of an entity they presume to be there but has not manifested physically. Group members will ask deliberate questions of an alleged spirit in a sequence followed by a gap of silence. ARIGs consider EVP collection to be scientific and objective because it uses equipment. A concerted effort is made in most ARIG ghost investigations to capture these recordings as evidence. It's been amply demonstrated by various EVP debunking efforts that ambient sounds the observer doesn't notice are recorded and subsequently misinterpreted as spirit communication. Any little sound in a quiet place can be caught on audio recordings—zippers zipping, whispers, sniffles, fabric rustling, and digestive noises. Because of this, many ARIGs will have strict investigation protocols about how to collect EVPs including rules against whispering, rustling clothing, and instructions on announcing who is speaking each time to clarify what is anomalous versus human-generated noise. The ARIG analysis of audio evidence consists of listening to hours of recordings made during an investigation to spot an anomalous sound. The playback may be manipulated by changing speed or enhancing sound. The faint audio clips typically require headphones and are distorted, low-volume or obscure. Modern EVPs have become less a curiosity and more threatening or frightening. The sounds are often interpreted as children, people crying for help or laughing, or aggressive entities saying "get out" or using obscenities. The presentation of EVPs on the websites always includes interpretation of the EVP, often given to the audience prior to listening to the clip. This action provides a leading explanation and eliminates objectivity. The human brain has an innate ability to make sense out of noise, to hear patterns and interpret the sounds as human voices. And, the listener will often hear what they are told they should hear.

The groups rarely verify their findings independently by testing the unprimed interpretation of the sounds and by seeking confirmation of measurements. Logically, if the voice is that of an entity at the location, multiple investigations should come up with the same voice or the same words. If the idea of spirit communication is correct, results at the same location should be compared and similarities should be found. Other additional equipment should be used to verify the recording.

EVPs are widely regarded by non-paranormalists as random sounds produced (unnoticed) by the participants, human voices recorded by accident, environmental sounds, stray radio frequencies, or machinery noise. Hoaxes can't be ruled out due to the uncontrolled conditions. Even if conditions for sound recording could be controlled, there are several other possible non-paranormal explanations that would require consideration before concluding that the sounds are generated from spirits. In my survey of sites, no ARIGs did anything beyond presenting various EVP clips as evidence. No ARIG provided a reasonable basis for the sounds to be from a discarnate being. While EVPs are curious things and fun for investigators to find, there is no evidence to support that they are more than various-sourced anomalies which the human brain interprets as meaningful. Interpreting these random sounds (or unnoticed voices) is an exercise in anomaly hunting and apophenia (Leary & Butler 2015). In stark contrast to lay investigators, modern parapsychology researchers don't focus on EVP and generally do not consider them as useful evidence.

Mystery Animal Sounds

Cryptozoological websites also provide audio as evidence of some living creature hidden out of sight. Recordings made by cryptid witnesses or investigators are claimed to be Bigfoots communicating through tree-knocking, booms, howls, screams, or speech-like chattering. The sounds, they say, are not identifiable as any known animal. Yet, this is not substantiated since it's not possible to know which commonly known animals are out there and the range of noises they might produce. With the ubiquity of

Bigfoot hunters assuming their generated calls will produce a response, the possibility is high that their knocking and howling are effecting a reply from another group on the adjacent ridge! Such sounds, as with EVPs, are presented at face value with subjective interpretations. In no case did I encounter an analysis by a sound engineer or linguistics analyst. The sounds are not identifiable by the investigator and, therefore, are assumed to be from the unknown entity, an unsupportable conclusion.

Physical Evidence

Physical evidence is that which has demonstrable physical characteristics such as objects, remains, or traces. Physical evidence is more substantial than testimonial, statistical, or analogical evidence because anyone can examine it for themselves. It can be objectively evaluated by experts or submitted for testing. ARIG cases don't often produce substantial physical evidence. The most famous evidence for paranormal explanations are photos and video, recorded sounds, footprints, hair, or effects on the natural surroundings (marks, burns, breakage, etc.). Physical evidence for paranormal claims is exceedingly rare or questionable but when such evidence *is* put forward, it is newsworthy. Pieces of extraterrestrial craft, alien implants, cryptid bodies or body parts, or ghostly ectoplasm that confirmed a paranormal identity would be extraordinary finds. However, none have turned out to be convincing.

Animal Traces and DNA

Field reports for cryptozoological investigations may contain mention of animal traces or remains, footprints, handprints, broken or manipulated plant material, trampled areas, smells, sounds, and observations of movement. Cryptozoologists strive to collect *any* potential physical traces including hair, feces, saliva, and partially eaten food as evidence. Plaster casts will be taken of any footprints or other suspected body part imprints. There have been hundreds of reports of Bigfoot footprints, hand and knuckle prints, and even a buttock print.

Hair strands caught on branches and fences as an animal passed are retrieved and preserved for attempted identification. Food bait may be used and, if partially eaten, the very serious cryptid hunters may preserve the food item for DNA testing of the saliva or to analyze bite marks. Livestock carcasses may be taken to an expert to determine if the prey was nabbed by a typical predator or something atypical.

In the case of supposed chupacabras (a vampire-like "goat sucker" that is said to kill livestock) in the southern U.S., several of the supposed creatures were trapped or shot and preserved to be examined. When biologists and veterinarians examined the creatures, and sometimes analyzed the DNA, the very strange-looking animals turned out to be nothing monstrous, but were canids or native animals with a skin disease called mange or some other hairless condition that caused fur loss. Even these solid conclusions based on the best physical evidence that can be obtained in cryptozoology, a specimen, do not deter some amateurs from belief that a mystery animal is still to be verified. These hairless canids have become known as "blue dogs" by researchers who think they are something new in nature.

Though hair and remains containing DNA are sought for evidence, they are not a slam dunk. The analysis of biological data like hair or DNA typically returns a verdict of "inconclusive" because not enough information is available to make the determination of a match to a known animal. The DNA may be degraded, improperly collected, or badly preserved. Hairs are difficult to match conclusively. If the result is inconclusive, or "unknown," ARIGs sometimes frame this as proof of an "unknown" creature. Within the past few years, DNA evidence collection has become more prominent as the cost of genetic sequencing has lowered. Blood, feces, hair, nail and bone samples have been collected for DNA testing.

UFO Traces and Remains

It is very rare for UFO reports to have any physical evidence associated with them. The exceptions include burn marks, manipulated vegetation, or mystery substances composed of threads or gel/slime. Some abductees claim to retain physical scars and marks, even implants. The belief that aliens are implanting tracking devices into humans is a fringe belief to which a few UFO researchers subscribe. Suspected implants have been revealed to be man-made materials either accidentally or purposely embedded. Claims that they are made from *strange materials that are not of this earth* have never been substantiated. Most chemical laboratories would be able to ascertain if a sample didn't match that of earth materials, but none have reported such a finding.

Ectoplasm

During the era of spiritualist mediums, a form of ideal physical evidence was regularly produced: *ectoplasm*, a wispy or gooey substance that came out of the mediums' mouths, noses or other places (use your imagination). Sometimes the ectoplasm formed phantom limbs or resolved into the faces of the departed. Ectoplasm was said to be the means the spirit used to manifest itself. Davies (2007: 130) describes it: "Ectoplasm was thought to be the solid essence of mesmeric fluidic forces, moulded by the sympathetic energy generated between the medium and the spirit world, allowing the dead to manifest themselves physically. Proof, at last, that ghosts had substance!" Observers were instructed not to touch ectoplasm or the apparitions in the dark séance room lest the medium be spiritually injured. Production of ectoplasm was eventually concluded to be a parlor trick mastered by mostly female mediums. Ectoplasm back then looked like (and later turned out to be) thin fabric, gauze, paper, or cheesecloth. It could also be string, bits of rubber, egg whites, or animal parts (some mediums got creative) sometimes glowing from luminous paint. The mediums would hide the substance or would swallow and regurgitate it in a dramatic appearance during the performance.

With the advent of *Ghostbusters*, ectoplasm changed to green slime. Regardless of what sort of residue a revenant leaves behind, it would behoove the researcher to collect a sample. Alas, in today's world, finding paranormal ectoplasm is a rarity. Any collected material would require testing by a qualified laboratory, preferably more than one. Only then can a researcher make conclusions about its origin. Ghosts leave a slime trail in fictional tales but it makes no sense for non-material entities to leave physical traces. We are left with essentially no physical evidence of ghosts and whatever other physical effects have been suggested (such as injuries or marks on the environment) are of questionable origin and authenticity.

Injuries and Bad Feelings

Ghost investigators find very little in terms of physical evidence to use in their research. In rare cases, but becoming more commonplace, investigators and those who live in a reportedly haunted location will cite scratches or marks on skin, walls, or objects that appear during the investigation or overnight as examples of physical encounters with an entity. Smells, feelings, or physical changes to the environment are also reported but rarely confirmed and investigated. Such effects are transitory and difficult to record objectively. Or they could have always been there and just noticed upon this new survey. Injuries discovered during an investigation cannot logically be attributed to entities when there are many alternative, non-paranormal explanations including self-infliction.

A frequent claim cited in a supposedly haunted location is an unusual "feeling" experienced by the researcher. Perceptions commonly reported in association with paranormal activity include coldness, chills, nausea, hair standing on end or a static electricity effect, sadness, dread or fear, difficulty breathing or chest heaviness, and a sense of being watched. In some cases, investigators will report being touched, slapped, poked, or pushed by something unseen. All these perceived feelings can be produced by the effects of expectation, fear, anxiety, and excitement. Or, they may be made up in the moment to gain attention or to add drama. The observations can neither be confirmed nor are they helpful to those of us who were not experiencing the same stimulus. Their claims are subjective and basically worthless without additional supporting, objective evidence. Ironically, they are the most persuasive evidence to the individual who felt them. They trust their own interpretation which tends to become stronger and more elaborate as time passes and the story is related in the framework of paranormal belief.

Apparitions

People have reported apparitions for over 1,000 years—featuring dead warriors, white sheets (burial shrouds), moaning, and chains. In a sense, not much changed, but we note that chains are not very common anymore. For a complete history of ghosts and their evolution from ancient to modern, see Finucane (1996) and Davies (2007). To generalize the history of ghostly apparitions, we can fairly say that they reflect their times.

An apparition is the appearance of a person, animal, or thing in a manifestation considered to be supernatural or at least paranormal. They can represent a person or thing from the past or from the present but in a distant location. Pop culture has given us frighteningly realistic, horrible, and highly creative apparitions. In society, reported apparitions are described as glowing figures, shadows, clouds, or mists. They may be transparent, or completely solid like a regular person. ARIGs report encountering apparitions and will attempt to photograph or record them. Unfortunately, cameras capture light anomalies, glitches, cigarette smoke, condensation clouds, or actual objects that are then called apparitions.

Some historic places are said to be haunted by a famous past resident whose apparition is regularly and consistently reported. It would make sense to document reports of apparitions to see if there is a pattern associated with a place or time. For example, an apparition that appears on an anniversary date is a testable claim, but that requires an assumption that the tales are accurate and truthful. Many are exaggerations. Tales of apparitions are popular stories and many people visit a location intent on describing any anomaly as a ghost to become part of their legend-tripping. Testing the regular appearance of apparitions would require controls—all witnesses would have to be blinded to the past information about the location and reports. We would need to assure that there was no leakage of information, and that each witness reported the experience independently, objectively, and accurately.

Though apparition encounters are greatly desired by ARIGs (and skeptics) as they are powerful, convincing personal experiences, apparitions leave no physical traces (except in movies). We can consider them as curious but they do not tell us much of anything.

Rock Throwing

Rock-throwing incidents have been described throughout history. Reports of rock-throwing (from tiny gravel to heavy blocks) are associated with hauntings, poltergeists (interpreted variously as ghosts, demons, or psychokinetic powers) (Houran & Lange 2001) and even Bigfoot encounters.[5] Long ago, incidents of rock showers inside and outside a building were seen as a sign of demons or witchcraft. Centuries ago, rock showers tormented families in their village houses. Today, rock-throwing is reported rarely but currently in the context of Bigfoot encounters as a means of intimidation. ARIGs report the rocks almost never hit people. In spite of it being seemingly easy to trace the trajectory of the object thrown, rock-throwing incidents are still reported as inexplicable events with an indeterminate, and assumed mysterious, cause.

Atmospheric Conditions

Ghost/ARIGs often provided a graphic on their websites that linked to a data feed reporting "space weather." This included the moon phase and the state of geomagnetic activity in the atmosphere. About 40% of groups consider the moon phase or solar activity in their investigations (Duffy 2012). This belief is related to the idea that paranormal activity is more prevalent during times of active solar storms or during the full or new moon. However, the reasoning behind the information is almost never explained or referenced, nor has it been sufficiently demonstrated to be correlated. It appears to relate back to the concept of electricity or similar "energy" as a conveyor or enabler of paranormal activity.

Several scientific paranormal studies have investigated effects of magnetic fields. Michael Persinger's work is often cited as evidence that low-intensity, complex magnetic fields may induce a sense of a presence. However, Persinger's tests were done in a laboratory with controlled fields. Monitoring such fields requires refined equipment that amateurs do not have access too. In addition, our normal environments are flooded with electromagnetic fields from any electric-powered device. Contrary to the frequent mention by paranormal researchers, published research has not shown a clear correlation between magnetic fields and paranormal experiences. People react differently, often not at all. Perceiving a sense of presence is not the same as experiencing a ghost. Priming effects may be more influential than any electromagnetic fields. A few studies suggest that a rapidly changing field around a metallic object such as a bed frame may affect perception,[6] or they claim EMFs from bedside devices such as clock radios may produce an influence.[7] Richard Wiseman conducted a study in Hampton Court Palace in Britain[8] to see if paranormal experiences or uneasiness were correlated to magnetic fields. The research results were not definitive (Maher 2015).

Per parapsychological research, correlations between psychic activity, psychokinesis effects and ghost reports with geomagnetic and electromagnetic fields are not straightforward, have not been reliably reproduced, do not have a mechanism by which we can explain how they work, and are generally contentious. Yet, ghost/ARIGs erroneously repeat these statements as fact. (See Chapter 14 in Cardeña 2015.) Unfortunately, not only are foundations of these ideas unfounded and unreliable but there is internal misunderstanding and misinterpretation of the results between controlled experiments and the complicated real-world environment.

Occult and Religious Practices

The search for ghosts as evidence of an afterlife is steeped in metaphysical, spiritual, and religious contexts. Even Bigfoot is thought by some to be a spirit of the forest who can be worshipped or brought close through native rituals. We've made collective attempts to contact extraterrestrials by sending positive mental messages *en masse*. Use of occult means to obtain evidence is predicated on the belief in that these entities exist and can be detected.

Many ARIGs, especially ghost/ARIGs, employ several non-scientific practices alongside what they consider to be scientific methods creating a noticeably stark juxtaposition between science and anti-science ideas for those groups who claimed to be scientific. Their methods include a mix of equipment to detect environmental metrics along with dowsing rods, pendulums, Ouija boards, religious paraphernalia, charms, occult and religious rituals, numerology methods, and other New Age concepts. Groups that explicitly promote their scientificity justify their wide range of procedures by suggesting that there is no harm in utilizing these alternate ways of knowing. But they downplay their significance or state that they only use them as guidance to point out locations in which to use the scientific equipment. Unfortunately, this gives the appearance of throwing everything at the wall and "seeing what sticks." Some groups, though, will outright reject these "other" methods as a way of enhancing their credibility and emphasizing their sciencey-sounding methods.

Psychics

Use of psychic mediums has become increasingly more common, likely due to the many psychic-themed reality television shows and celebrity psychics who claim to communicate with dead relatives. Ghost-hunting groups typically have one or more members that claim they are "psychic," "clairvoyant," "sensitive," "intuitive," or even serve as "universal catalysts" to "assist investigators towards a scientific conclusion" (S. Hill 2010). Psychic aspects are also mentioned in reference to Bigfoot and the more supernaturally inclined cryptids as well as aliens who are said by some witnesses to communicate telepathically. The ability of one ARIG member to claim to gather information from psychic sensitivity sets up a potential conflict with the more skeptical member(s). *She* is the "feeling" while *he* is the "thinking" person. The "feeling" person is assumed to be less of a threat to another psychical being and thus have a greater chance to effect meaningful communication. The non-psychic, unable to verify these feelings, is useless.

As with psychic "readings," detailed information of psychic communication with spirits, like names and specific dates, are is rarely provided. If such verifiable facts are given, they are not typically checked. In no instance did I find psychic information provided that was independently correlated with verified information on the site. The "intuitive" information is vague, often dramatic, or open to interpretation. ARIGs do not conduct a test for validity of psychics by asking multiple psychics, blinded to a situation and to each other, to provide a reading at a location. This reasonable test should produce the same type of information from the same location if indeed they were utilizing paranormal senses. I have not found that such a test was done. Instead, psychic contributions are wrapped into the narrative of the investigation as they are dramatic and useful, but are discarded and ignored if they impede the paranormal explanation.

Ouija

The Ouija board is a controversial tool in paranormal circles with paranormal proponents highly polarized over its use. Some consider the

Ouija board (sold toy as a children's toy) a dangerous device that increases troublesome paranormal activity and opens "the portal to demonic entities" or "doorways" for (presumably evil) spirits to enter. Those who are invested in the spiritual aspect will warn against its use for these reasons. Or some ARIGs reject its use because of the irrational, subjective nature of it.[9] During my website review, several ARIGs hosted a "No Ouija" badge on their web site expressing their opinion. However, their reasoning for the objection was not obvious in all cases. As with the use of psychics, a few ghost/ARIGs who state they are scientific will still use the Ouija board as a form of provoking spirit communication, encouraging them to appear by talking or calling to the entities. The anti–Ouija camp will consider this to be provocation and discourages the practice. Less provocative objects are also used. About a third of ghost/ARIGs employ a trigger object such as a toy to entice the local spirit to engage with the object. Some will yell or use music or noise that they think might be relatable to the spirit. If the investigators have some idea about the spirit they are trying to reach, they will say names and ask questions to invoke a response. While such methods of enticing responses could readily be made into a testable experiment, ARIGs have not taken on that task of showing the Ouija or any other provoking method does or doesn't do something.

Dowsing and Pendulums

Divination devices are frequently found in the ghost/ARIG's box of tools. Dowsing (or divining) rods are the most common as they are believed by some to be useful in locating spirit energy or environmental anomalies. They are sold at paranormal conferences to ARIG participants who seek out all types of strange energy. Pendulums are less frequently used but can be more decorative, attractive and "magical." The sway of the pendulum—left/right or front/back or in circles—is interpreted to mean different things. As with the Ouija board planchette or pointer, the movement of dowsing rods or pendulum is influenced by the user, usually without them being aware they are subtly influencing the device. Therefore, these devices are not objectively reliable and there is no scientific basis for their use in locating energy in the environment of any sort. A considerable body of research results have shown dowsing rods and pendulums are not accurate at finding any object or water under controlled testing conditions. ARIG members, when prompted, will explain that they do not know exactly how these divination devices work but that they do indeed work to locate spirit energy or UFO landing sites, a clear example of magical thinking that much of the public appears to subscribe as well.

Religion

Several ghost/ARIGs incorporate a component of spirituality in their methods or the explicitly described faith-based methods they typically used. About 40% of paranormal investigation groups will identify with some sort of religious doctrine with around 20% particularly seeking spiritual or religious knowledge as a goal of their work (Duffy 2012). This percentage likely reflects the general population interested in seeking spiritual answers. Twenty-four groups out of 1,000 studied were *explicitly* religious; that is, they were affiliated with a religious institution, run by clergy, or guided solely by religious principles. Most religious-based groups promote a Christian viewpoint but there are those who will respect and accommodate whatever beliefs the client holds. A few groups will use obvious religious-themed methods in conjunction with stated "scientific methods." These groups are not necessarily those affiliated with religious institutions or clergy (though some are). Several who state they are scientific and do not clearly exhibit religiosity, still invoke protection from demons and Satan in the form of incorporating their faith-based beliefs in their interpretations, wearing Christian symbols, or conducting group prayers prior to the investigation.

Discussion and invocation of demonology has definitively increased within the amateur paranormal field over the past decade. Several ARIGs include information about demonology on their sites. Multiple groups publicly declared a specialty in malevolent hauntings or demon infestation. One group created a specialty sub-group to address activity "when scientific methods have been exhausted." This group also stated they have been "trained" in such matters.[10] Being trained in demonology is a fairly frequent claim never accompanied by an explanation of how individuals are trained in this subject (that could reasonably be seen as a subset of either theology or folklore). Self-styled demonologists present the classification and characteristics of demons (assumed to be real) as careful, meticulous "science" (which explains the "-ology" suffix) and, in one case, is referred to as an "unconventional science."[11] According to Tea Krulos (2015) who tagged along with ghost investigation teams, it is not unusual to find people heavily invested in demons as the potential cause for their current troubles. Exorcisms have become somewhat trendy in the 21st century, and requests to Catholic exorcists have increased.[12] This may also be in response to media portrayals from *The Exorcist* to the many films and documentaries (fiction and non-fiction) that use demonic themes or blatant imagery. On Halloween, 2015, a live exorcism of a house was shown on the channel Destination America. Removing demons from a house sounds ridiculous and, apparently, nothing happened, but the spectacle still got decent ratings showing that people are curious and willing to watch.[13]

The demonology aspect in the hands of amateurs is a disturbing trend which can result in making residents in a home even more distraught by exploiting their belief in evil spirits or demons. Several ARIGs state they have attempted exorcisms of people or places—a serious and unethical practice since those who suffer from a torment labeled "demons" are in need of mental, medical, and behavioral professionals, not amateurs who may think they are equipped to face down a demon because they read some grimoire (textbook of magic). There is no question that the pursuit of this line of action—to suggest that a demon or the devil himself is the core cause of a disturbance—is arguably the worst grievance we can hold against ARIGs.

In contrast to those that advocate use of demonology, other ARIGs, especially most of those who advocate scientific methods, completely eschew horror imagery and any talk of demons.

New Age

Less mainstream religious or spiritualists beliefs are present in paranormal subject areas that may reveal themselves in ARIG discourse. The Spiritualist movement was the earliest effort to connect to those who have passed from physical earth. Spiritualism as a practiced religion is much less common today than it was at its peak in the 19th century, but it has not completely disappeared.

There are more than one UFO religion. UFOs have frequently been viewed as transport for ancient helpers, angels, perhaps even a futuristic time- or dimension-traveling technological race. Scientology, the secretive and frequently litigious religion that recruits Hollywood celebrities, founded by science-fiction writer L. Ron Hubbard under dubious circumstances, has space aliens as part of their teachings.[14] The Raëlians are another UFO-based religion that sometimes appear in the mainstream news for their outreach projects.[15]

There are no Bigfoot religions (yet) although Native Americans consider Sasquatch or something very much like it as a spiritual being or native "brother." Crypto/ARIGs frequently use Native American folklore and spiritual beliefs as evidence to support the presence of entities that people may be referring to as cryptids. Ideas of magical powers endowed to the creatures like Sasquatch or various shape shifters and tricksters are commonly discussed parallel to the consideration that cryptids are normal biological animals. With a strong cultural trend toward re-enchantment, to maintain and promote mystery and magic particularly related to the environment, personal well-being and feminism (A. Hill 2010), a Bigfoot/Sasquatch with minor supernatural powers could readily become a focus of adulation and even worship.

The depth and commitment ARIG members show towards supernatural causes and spiritual meaning of claims has not been measured. Interest in the topics can surely take one "down the rabbit hole" into unhealthy situations. People who treat these beliefs so seriously that they serve as a religion have no need of investigation into claims and remain apart from ARIG activities. But those who start by wondering and investigating claims, who then become heavily invested in belief, may slide down this slippery slope into obsession or worship. Undoubtedly there are those who become obsessed with the topics and allow paranormal beliefs to overtake their lives. Ghosts and demons may be perceived to be influencing the environment and directing people's lives. There are those who believe Bigfoots are in their midst and speak telepathically to them only (Sykes 2016). UFO-related religions are often deemed "cults" in that they require adherents to cut ties with their family and friends in order to preserve the association. The most notable UFO cult was that of the Heaven's Gate religious group who committed mass suicide in 1997. Thirty-nine people died based on a belief that extraterrestrials would visit, in conjunction with the Hale-Bopp comet, and release them from their earthly bodies. Such extreme behavior illustrates the ultimate danger in blind belief.

Native Beliefs

Everyone *knows* that ghosts are associated with ancient burial grounds—we are conditioned to believe so thanks to the impact made from fictional tales like *The Amityville Horror* and *Poltergeist*. A common motif for hauntings is that of the spirit who was buried in an inappropriate place or whose bones have been disturbed or desecrated. The land of the ancient burials, according to these legend, is cursed and no peace will come to those who reside on that land. Ghost hunters will gravitate towards these burial sites, graveyards, or sacred lands with the thought that restless spirits remain there. The evidence that such grounds are haunted is derived from folklore and psychic pronouncements, sources that are not exactly factual.

With a rich collection of gods, spirits, and rituals, Native American beliefs are often invoked in the U.S. to clear out the bad spirits of an area, especially if the area was known to be historic tribal grounds. Some ghost/ARIGs will burn herbs to eradicate "negative energy." Sage, cedar, or sweetgrass is bound into a bundle and lit with the smoke said to cleanse the area. Ritual "smudging" is conducted by carrying the smoking bundle or bowl around the house, location, or the person affected. Prayers and incantations may also be used. This ritual has become more popular recently because it provides a visual formality that appears profound and mysterious (similar to the use of sciencey gadgets) and it's is easy to do.

Folklore and mythology of native Americans is invoked most readily in the field of cryptozoology. As mentioned previously, Bigfoot and several other mystery animals are frequently described in the context of spirit or sacred animals from Native lore, a manifestation of evil, or an evil omen. Cryptozoology stories are particularly rife with connotations of myths and folk tales of the previous cultures that existed in the area. Caution is required when applying legends, myths, and sagas to modern eyewitness accounts and interpretations of a phenomenon. Just as the ARIGs are not trained in scientific methods, they also are not trained in anthropology and folklore studies. They may misappropriate a natural explanation for a purely fictional tale. Some native cultures don't distinguish between natural and supernatural creatures as modern Western culture might, and the outsider may fail to grasp the context that determines the real nature of these creatures. Cryptozoologist writers examining old tales will cherry-pick items from a culture that they can use and ignore the rest that doesn't correspond to their preferred conclusion. There is also a hazard in using ideas of pre-scientific ancient cultures literally to produce scientific theories as our approach to and knowledge of the natural world is monumentally different. It is too great a leap to conclude that, long ago, people interpreted nature as we do now. Stories contain motifs, distinguishing or dominant features, that are entirely fictional but occur across several cultures—such as a water guardian, dragon, or wild man. They were not and should not be considered objective documentations of natural history (Meurger 1989). Myths may mean different things to different people and what appears to be an eyewitness account may not represent natural facts. ARIGs should be cognizant of the lure and inappropriateness of cultural stories being applied as evidence for a claim whether that be in relation to demons, angels, spirits, cryptids or UFOs, but this realization is lacking.

Spirit Cleansing

Several Ghost/ARIGs may attempt to do more than just identify spirits' activity. They claim success in "helping spirits ascend to a higher plane,"[16] or in "communicating with the astral plane."[17] They will tell clients that they can help these spirits move on from the place where they are currently being troublesome. Cleansing or ridding the location of bothersome spirits is promoted by some ARIGs or is a specialty service performed by some groups while others will offer it in an apparent last attempt to make the client feel better (about 36 percent from Duffy 2012). It is not always clear if they believe that it really works or serves a placebo effect.

Groups may explicitly avoid any meddling with potentially "dark" spirits. Some state that even if they identify that you have a spirit, they will not perform any rituals to address the problem. Several paranormalists who believe in ghosts also believe a ghost or spirit can attach to them and follow them home causing trouble in their home and life. Therefore, they use spiritual protection or cleansing rituals on themselves.

11

Inquiry and Investigation

Most of us will conduct some form of investigation in life, an inquiry into answering a question about a choice we must make (health treatment, major purchase, financial investment) or to figure out what is going on and what to do about it (in the course of our jobs, finances, or everyday mysteries to solve such as what critter is getting into the trash at night.) It's impossible to inquire with a completely open mind because we will have some inkling of what we expect to find and perhaps lean towards a preference from the beginning (our hypothesis). Fair inquiry requires that we not declare the finding before the facts. Our destination should not be set before we even begin to examine where it might lead us. Investment in a preexisting conclusion is called *ideological conviction* (Pigliucci & Boudry 2013). This conviction acts as an anchor pinning the holder to a spot making it hard to move. It can affect those who attempt to incorporate science in their investigations. If mainstream science is not amenable to the object of their conviction, they discard the science instead of the conviction. Even some professional parapsychologists have suggested that the rules of science be changed rather than give up their conviction.[1]

ARIG's goals are unique, necessarily different from the researcher in a lab, private investigator, or crime detective. But that does not preclude use of a logical structure, sound methods, and ethics in their investigations. The following section is a deeper dig into the process of ARIG inquiry and investigation that reveals flaws in their system.

What Is "Investigation"?

The fundamentals of investigation are that the investigator consider the standard *what, who, when, where, how* and *why* (Baker & Nickell 1992). The framework of the investigation should lead to determining some reasonable, defendable outcome or explanation.

A paranormal investigation should, logically, have the goal to understand what, if anything, has happened in a situation where a person or persons perceives to have experienced an event outside of typical or normal human conditions—whether that be a haunting, an observation of an anomalous aerial object, an unknown creature, or some other seemingly unexplained natural phenomena. "If anything"[2] is a crucial caveat that is often completely overlooked in ARIG investigations. Consideration of the probabilities of misperception, misinterpretation, and hoaxing or trickery, which are high, is essential since there is little sense in pursuing inquiry into what will amount to a dead end. That is, unless the goal is not to find out the most likely answer but to pursue some other end. What some ARIGs do might be a form of "phenomenology"—a discipline in philosophy that studies the subjective experience (Jenzen & Munt 2013). This is not concerned with the innate truth but just of the perception.

Bias

In research, it is critical to understand the effect of the person doing the observing on the observations themselves. A main tenet of science is to remove the bias (or preconceived tendencies) of the observer during the research. One means by which to do this is called "blinding." A party in the experiment—the subject or the experimenter—is deliberately left unaware of some aspect. This minimizes the effect of unintentional selections or preferences towards the outcome. If both the experimenter and the subject are blinded, the experiment is called "double-blinded." However, double-blinding isn't practical in real world observations and experiments, only under controlled lab conditions. But one example of "blinding" that could be used by ARIGs is to *not* know all the details about a paranormal event and to see if their unbiased collection of evidence supports the reports from witnesses. This is almost never done.

Another form of bias that is common to scientists and non-scientists alike is to seek confirming evidence that fits with your preferred conclusion and to discard that which does not. This is sometimes called "cherry-picking," which is collecting support for your argument and ignoring non-supporting facts. Valid critiques can also be ignored and disconfirming evidence can be explained away. If no standards exist in your field (such as with ufology, cryptozoology and paranormal investigation), then any anomaly can be fitted into the inquiry as evidence. The presentation of such cases may look convincing to those of the same opinion but will look weak to skeptical critics.

Most importantly, bias is inherent in human observation, even in trained observers. Therefore, people do not make the ideal instruments for observing and recording. ARIG participants are mostly believers in the mysteries they seek. Possibly without being aware of this belief bias, they tend to seek out the evidence that suits their preconceived notion of the cause. They *want* to find evidence for the paranormal. In fact, it's often their explicitly stated goal. This leaning, belief in the paranormal, colors the entire process of investigation, and can be cited as a key reason ARIG research is not scientific.

Beginning an investigation with belief and a hidden or publicly acknowledged desire to support that belief, will result in motivated reasoning—the investigator will seek out specific evidence—possibly engaging in anomaly hunting—to fit into their mental model of what they believe is going on. Anything that doesn't support the belief will likely be ignored or rejected. Motivated reasoning results in missed consideration of what may be important evidence or possibilities. ARIG members (although this happens to everyone) usually enter a situation primed (the opposite of *blinded*) to have an experience and knowing what others have reported. If a place is supposedly a "hot spot" of activity, every sound, smell, sight and feeling will be deemed significant and related to the preferred cause. Patterns will be seen in randomness. ARIG researchers display a tendency to group together anomalies, events, facts and clues that may actually not have a common cause. When you assume that the thing you are trying to find exists before you find it, you will "see" signs of it everywhere according to your own perception. This selectivity of observations—to pick those that conform to your beliefs but discard or disregard those that do not—is called *confirmation bias* and we all do it to some degree. Almost all ARIGs (with the exception of the rare few groups that practice scientific skepticism) assume that one of the possible, even probable, explanations must be attributable to their paranormal subject. They will accept weak evidence that would not be acceptable to the scientific community. They conduct activities as if subject X is real when these conclusions have never been established as knowledge. Instead of collecting evidence and making a conclusion, the conclusion is the frame onto which any evidence (anomalies) is hung. This is backwards and will lead to major errors such as false positives. A false positive is a piece of evidence (image, movie, audio recording, etc.) that has a normal explanation but is attributed to be paranormal. False positives are readily found in images and sound recordings without adequate consideration of alternate plausible explanation. Typical examples of these types of misrepresented evidence are EVPs, orbs, mists, shadows, footprints, disturbed vegetation, or indistinct objects. Mystery animal photos are almost always too indistinct or lacking appropriate scale to discern what the creature might be. In the case of Bigfoot, terrible photographs or videos are so common, they have their own moniker, "blobsquatches." About 90 percent of Bigfoot imagery might be called blobsquatches—you can't tell what it is. (I don't know, therefore, it's Bigfoot) Most blobsquatches are obviously ridiculous. But the person who produced it may be serious. They may truly believe the anomaly represents a monster. The same phenomenon happens as objects in the sky turn into UFOs and smudges on reflective surfaces become spirits.

Approaching Investigation with a Pre-Existing Notion

The obvious difference between the investigatory approach of a paranormalist (willing to include paranormal causes as an explanation) and that of a non-paranormalists (assumes you can't ever conclude anything is "paranormal") is an (intentional or unintentional) effort to look for or focus on evidence to "prove" a preconceived notion. The non-paranormalist approach is wider at the beginning, with less assumptions about the scenario. It starts with the questions "What really happened? Was there a mysterious event that occurred?" and only then will they pursue an answer to "What are the most satisfactory, normal explanations that may account for the events?" The non-paranormalist approach was not apparent as the starting points for most ARIG investigations. The process employed by many ARIGs is to take witness claims or circulated stories at face value, assume something is going on, and seek evidence to support those claims. As I described in my own ghost hunting experience at Fort Mifflin, paranormalists were not amenable to my approach that was more aggressive towards establishing "what, if anything, is happening" and we clashed. They were disappointed with my suggested rational explanations and rejected them. This was not unexpected as it is unlikely for people to reevaluate their observations and conclusions if those conclusions are supportive of their preferred worldview. It will not help to point this out, either. ARIGs will instead become defensive in the face of skeptical criticism to protect their worldview.

Ruling Out Causes

Ruling out natural causes is a basic expectation in any paranormal investigation. ARIGs can do an adequate job considering the time they have to explore these options. However, many don't spend enough time or effort to do so completely. They are disadvantaged in finding the precise non-paranormal cause in that they do not have the luxury of spending weeks or months at one location or on one event to study it thoroughly, as a scientific investigation would be designed.

Attitudes and practices of the investigator are key to the results. An investigation is "better or worse conducted depending on how scrupulous, how honest, how imaginative, how thorough it is" (Haack 2007: 339). Not considering the many alternate explanatory causes for a phenomenon is a trap for ARIGs as they fail to develop and nurture a critical attitude (Beveridge 1957; Baker & Nickell 1992; Radford 2010). ARIG's frequently will state as part of their methodology that they pursue all avenues of the logical or normal first. Only when all those options have been exhausted, do they conclude that something other than the normal exists here. Such statements are unreasonable. How is it possible to think of, test, and rule in or out all possible normal explanations? It takes a great deal of time, effort, and imagination to think of and devise testing for the many logical possibilities, but we can't ever know all the potential possibilities.

But the main reason why the normal causes are not diligently sought appears to be because of the bias in favor of a paranormal idea. Thus, the context is conducive to finding evidence that will be consistent with that goal. Once that pro-paranormal evidence is found, they often conclude the investigation. Collecting further data or assessing validity of the evidence is irrelevant because a satisfying explanation has been reached. The effect of the bias towards paranormal causes leads ARIGs to mistakenly reinforce their preconception that they have found another supporting case for paranormality.

Skeptical Investigators

In the 1957 volume *The Art of Scientific Investigation,* Beveridge says, "Nothing could be more damaging to science than the abandonment of the critical attitude and its replacement by too ready acceptance of a hypothesis put forward on slender evidence." Anyone who would embark on investigations with a preconceived idea of the cause of the mystery is not doing good science or competent research. A few truly skeptical investigators who pursue solving these cases via a rational approach, outright state they hold no prior belief in any paranormal causes. From this basis of critical thinking, we can contrast the different methods of approach to an investigation. Even though there are a significant number of skeptically-minded organizations in the U.S., few publicly promote paranormal investigation as one of their services. Presuming paranormal activity has nearly become a prerequisite for ARIG participation. And, skeptical investigators would not be convinced by the typical evidence presented in eyewitness accounts. Only two of the groups in my survey explicitly identified themselves as skeptical organizations.

Sham Inquiry

Proper research is a well-organized, continuous action that builds a foundation culminating in progress in understanding the subject. Inquiry into the subject as part of the overall research agenda must have a clear purpose, with a goal and a plan to achieve that purpose. Philosopher Susan Haack (1997) brought forth the idea of *sham inquiry* ("fake" inquiry or "pseudo-inquiry") via the musings of C.S. Peirce regarding *sham reasoning* (Peirce 1931). Sham inquiry is not an often-used term but it resonated with me with regards to the method used by television paranormal investigations and, subsequently, ARIGs, to investigate mysterious claims. Sham inquiry, is a backwards form of inquiry where, instead of the evidence leading one to a conclusion, the assumed conclusion determines what the reasoning shall be. That is, the inquirer already has a preconceived notion of what he wishes to conclude. This method is practical for a television program or to impress a client. It is also excellent for reinforcing an existing worldview. However, the "sham" process means going through the motions, without understanding or foundation, and placing adherence to a belief framework before truth or accuracy. The process and the results of it misleads the investigator and misinforms the audience—it's dishonest. Sham inquiry puts us several steps backwards as it allows the investigator to make an end run around adequate consideration of important questions. And, it speciously portrays an unsupported presumption as the reasonable conclusion to a research inquiry.

After extensively examining their methods, and documenting ARIGs' thinking about what investigation and scientific inquiry is, I concluded what most ARIGs do can fall under this label of sham inquiry. Evidence provided here shows that ARIG investigation often turn out to be an exercise in advocacy of a personal belief. The outcome of decades of unproductive and ineffectual sham inquiry is that society has no more reliable or better quality information on these paranormal subjects than we had 100 years ago.

What follows are some examples taken directly from ARIG websites and publications that illustrate how many ARIGs misunderstand the scientific endeavor and undertake sham inquiry while claiming they are being scientific. Groups may claim to be "unbiased," only to then state later down the page that they seek to "validate the existence" of their quarry. Those two concepts are contradictory. If you are seeking to validate the existence of, say, Bigfoot, then you are not unbiased in your search—you are seeking to find Bigfoot, not an alternative answer.

An illustration of this comes from the bylaws for Milwaukee Area Paranormal (MAPI). The website stated:

> The purpose and objective of this organization shall be:
> A. To scientifically and without bias or prejudice explore the realm of the paranormal.
> B. To attempt to prove the existence of claimed paranormal activity or beings.
> ...
> E. To educate the membership and public on the existence of the paranormal.

Stating in A that they are without bias is contradictory to their B and E values which make it clear they are biased in their belief of the paranormal and intend to present this to the public as true. Another example[3] seems to provide only lip service to normal explanations with the intent to reinforce a preconceived paranormal belief: "A paranormal investigator will rule out any natural causes ... and then pursues the paranormal side of events. This ensures that the evidence collected can be proven without a shadow of a doubt that the events recorded are in fact paranormal and ghostly in nature." Unpacking this statement requires us to question how the "paranormal side" can be pursued when the "paranormal" is exclusionary and unexplained. We see some words of certainty, "proven without a shadow of doubt," that no scientists would use in a project. How can you prove something is "ghostly in nature"?

These two statements from groups betray a perverse understanding of scientific philosophy and methodology: "We cannot be so vain as to rule out that which is only scientific in nature"[4] and "There are times when we are left with evidence that proves science has no understanding."[5] Some groups explicitly express a desire to change science to suit their paranormal ideas. They wish to "adapt existing science laws to the reports of the paranormal"[6] and "attempt a bridge between science and the paranormal."[7] The Ohio Paranormal Researchers redefined paranormal in their own terms: "We make conclusions by using our own discription [sic] of what we think is paranormal ... which is not anything that is readily explainable by known scientific methods." Science is a community effort, an ethos, a method, and a body of knowledge, not a set of statements written in stone. Anyone who has some scientific training or familiarity with scientific discourse likely found the above examples painful to read and laments the degree of scientific misunderstanding in the U.S.

12

Pseudoscience

The ARIGs we've been discussing throughout this book uniquely focus on those areas where no other organized research or inquiry is focused—those on the questionable fringes of experience, or "paranormal" activity. ARIGs may respect science but they misunderstand it. This misunderstanding is evident in their ubiquitous conclusion that they have found evidence or proof of the paranormal, in their methods of investigation, and in their philosophy towards investigation. This section wraps together all the observations about ARIGs to assess if their work constitutes "false science" (pseudoscience).

The scientific community generally steers clear of paranormal subject areas but will disagree about the importance of these topics and possibly even be hostile towards those that investigate them. As established, a presentation that *looks* scientific can increase prestige and credibility with an audience. Therefore, even those with religious agendas will attempt to use science when possible to raise their status. Scientists will not hesitate to protect what they feel is their turf—explaining nature. Therefore, scientists have historically constructed social and philosophical boundaries around what they deem to be science and what to exclude. This has not been particularly successful for reasons I will attempt to elucidate.

ARIG websites can, unfortunately, be exhibits of sloppy thinking. Problems displayed by ARIGs and individual paranormal proponents include ignorance of normal human biases we all have, lack of critical thinking, contamination from belief, misappropriation of skepticism, influence from media sources, and lack of adequate background knowledge. Their results, however, are packaged to compete against natural, scientific explanations. Thus, cryptozoology, paranormal investigation, and ufology are commonly labeled *pseudoscience*.

The Demarcation Problem

When scientific-minded commentators think a field of study claiming to be scientific is on the wrong track, they make no bones about it—it's labeled "pseudoscience." That's a damning word. The label serves as an efficient rhetorical means to represent an inferior alternative to science. "Pseudoscience" entered wide use after the 1960s (Thurs 2007). Its first use describes a "pseudoscience" as having a "self-adjusting arrangement" whereby the evidence was either kept or discarded in accordance with preserving the idea at all cost (Pigliucci & Boudry 2013). The growing use of the word suggests that science, as an institution and cultural icon, is policing the boundary between what is acceptable science and what is not (Thurs 2007). Fringe topics are frequently labeled as pseudoscience (Friedlander 1995; Hines, 2003): ufology, "Bigfootology," cryptozoology, creation science, and "ghostology" are all examples that have been soundly tossed into the bin of pseudoscience and excluded from academic circles.

Literally, *pseudoscience* means "false science" and it is employed effectively as a value-laden pejorative to indicate exclusion from the traditional realm of science, and, consequently, from legitimacy (Haack 2007). However, it's tricky to use this way because there is a quality spectrum ranging from poorly done science, through dubious methods, to conventional, solid processes, to those that may be legitimate but new and unusual (Lyons 2009). So where can we draw lines, if at all, around bad science, non-science, pretend science, proto-science, and marginal science? Well, lines can't easily be drawn. The trouble with demarcating science from methods and knowledge generation that is not science is called the *demarcation problem* (Gieryn 1983). Adding to the cloudiness is that some subject fields go through maturation stages where they can't be fairly judged as to their place within the scientific spectrum. Designating a field or subject as pseudoscientific can be complicated and arguable. A field may be pseudoscientific in one era only to undergo a renaissance and become an accepted science, such as alchemy that evolved into the scientific field of chemistry. It's not the academic standing of any field (or person) that indicates pseudoscience (or pseudoscientist) but the method used and the type of knowledge produced. Lyons (2009) provides several examples, including that of sea serpents and spiritualism, to show that "in the midst of discovery, it is often difficult to distinguish what constitutes science from what does not." Some fields go off the track into pseudoscience where others eventually are legitimized. There is a spectrum of pseudoscience—from fields or individuals that just miss an arbitrary level of scientific credibility all the way to charlatans and medical quacks who deliberately deceive the public with sciencey-sounding tactics they know do not or cannot work.

There is always the possibility that breakthroughs in technology or theory, over time, can render a field previously labeled "pseudoscience" as legitimate. An example includes the idea that rocks fall from space (meteorites) (Westrum 1978). Until sufficient evidence was accumulated, the intellectual minds of the day did not accept that meteorites were space rocks. String theory has some characteristics of pseudoscience right now but because it's studied by qualified researchers building on what is already known about the universe; it's not typically called pseudoscience. ARIGs in my study frequently contend that theirs is a proto-science, one day to become legitimized. However, they fail to consider that each subject once was examined in terms of science and then rejected or evolved into another field. Such fields were not doomed to remain trapped in the morass of pseudoscientific wheel-spinning, but the odds that paranormal subjects as defined by ARIGs will undergo a reboot and scientific blossoming after all these decades of trying are remote.

Pseudoscience can not simply be characterized by a single trait but is more meaningfully described as a set of cumulative characteristics (from the scientific point of view) (Bunge 1984; Derksen 1993; Dolby 1979; Hines 2003; Pigliucci 2010). The more of the following characteristics that can be attributed to a stated doctrine (e.g. Intelligent Design) or a field (e.g. cryptozoology) or kind of activity (e.g. ghost hunting), the greater the chance it will be labeled pseudoscientific by mainstream scientists. Here is a list of key characteristics that contribute to being labeled "pseudoscientific":

1. Portrayed as being scientific by use of jargon, images and actions associated with scientific work;
2. Supported mainly by belief or problematic, weak or nonexistent evidence instead of sound, accepted research that would be convincing to mainstream scientists;
3. Lacking coherent, progressive, explanatory theories, or structured to be irrefutable or unfalsifiable;
4. Lacking internal critique or organized skepticism; questioning is unwelcome;
5. Proponents exhibit paranoia and feel a sense of persecution from mainstream science. They embrace unorthodox ideas and considering certain leaders as mavericks, even to the point of comparing themselves to Galileo;
6. Lacking interaction and overlap of research with other cognitive fields. No cumulative results are published and no real progress can be noted;

7. Proposal of unreal or not certifiably real entities and processes that are not logical.

In terms of *methods* that would be considered criteria for labeling a field or a claim as "pseudoscientific," the following characteristics are indicative:

1. Lax rules for data collection and experiments;
2. Lack of adequate environmental or experimental controls in positive studies;
3. Methods of research or evidence collection are haphazard, conceptually unsound or flawed; or, no research or active inquiry is being conducted to support the theory;
4. Heavy emphasis on soft data such as anecdotes and subjective feelings; use of unconventional, defective, or baseless procedures;
5. Use of special pleading to explain validity of results or shifting of the burden of proof. (Example: Skepticism creates negative energy and thus prohibits good results.

Any of these characteristics signal something is amiss within the field or methods. I hope I've made clear the good reasons why science rules are strict and difficult to attain. Therefore, when a field does *not* attain those standards, the results are not as reliable. If a field or idea strays too far "from the epistemic desiderata of science by a sufficiently wide margin while being touted as scientific by its advocates, it is justifiably branded as pseudoscience" (Pigliucci & Boudry 2013). Proclaiming something is pseudoscience creates an immediate contrast with science. When addressing claims about nature, however, a more general consideration might be simply to know if and why certain claims are more acceptable than others in terms of reliable knowledge. Unlike science, pseudoscience lacks a unity of practice and integration with other fields. It just doesn't make sense based on what we have established about the world.

For decades, the academic field of parapsychology has been fending off attacks from those who label it pseudoscience. Edzel Cardeña, a professor of psychology at Lund University in Sweden and director of the Centre for Research on Consciousness and Anomalous Psychology, insists that it is not. In the 2015 book *Parapsychology in the 21st Century,* he asserts that parapsychology *is* scientific and the rest of the 400+ page collection of articles are cited to support this. Cardeña, though, admits that amateurs currently using some parapsychological concepts (ARIGs) are far removed from the state of professional parapsychology (Cardeña 2015a). Parapsychology, however, suffers a lack of investment in the ethos of science in that researchers wish to claim special conditions, allow softer data when the laboratory experiments can't be replicated, and claim effects from "implicit psi" which means the experimenter or any participant is unknowingly affecting the results. There is no known way to shield an experiment from consciousness, especially if psi works retroactively (backwards in time) making some claims possibly untestable. More concerning is the attitude of modern parapsychology leaders to eschew materialism. Science is necessarily void of spiritual meaning, relying on a foundation that all things are made of matter and are to be explained only in that framework. Thus, the inclusion of non-materialistic and non-naturalistic ideas into the realm of science will not be tolerated within the boundary.

Is What ARIGs Do Pseudoscience?

Like science, "pseudoscience" is used to describe both a process and a body of knowledge. Can we also apply it to a community? Are ARIGs part of that community? Looking at the fields of study and methods used by ARIGs, we can point out the presence of many characteristics of pseudoscience. Within ghost "science," cryptozoology, or ufology, we immediately notice a science-like framing without any standard framework for evaluating the many explanatory ideas and speculation put forward or any formal process for considering them. Those who claim their ideas are scientific are not required to put those ideas to an organized community for testing; therefore, there is no formal system of error correction (Pigliucci & Boudry 2013) and no cohesive process to obtaining knowledge. Ideas are propagated even in the face of critical dismantling by others and production of contradictory evidence.

Since most ARIG members are not trained in science, poor scientific methodology is not surprising. Even credentialed scientists can fall into pseudoscientific practices if not careful. When the impetus behind propping up an idea is the strong preference for something to be true, even PhDs will find themselves accused of pseudoscientific practices to justify Bigfoot, life after death, etc.

Media Influence

Two factors that allow for pseudoscience to be smoothly sold to the public as legitimate are: (1) a spokesperson who displays some real or faux scientific credentials (thus, gaining social authority), and (2) a receiving audience with lack of understanding about what makes something suitably "scientific" and reliable. The media enhances factor #1 by covering questionable work as if it was science, often in a misplaced attempt to journalistically balance opposing viewpoints. For example, a global warming denier may be a meteorologist (weather forecaster) which suggests valid credentials for their position. But, only a very small number of scientists of all kinds, but especially climate scientists (different discipline than meteorology), subscribe to that idea. The depiction of one opposing view and one non-opposing view when there are, in fact, dozens of non-opposing views, is not an accurate representation. Evidence for climate change, or evolution, or the age and shape of the earth, or any well-supported scientific tenet, is substantial but deniers will always exist. Viewpoints disputing those tenets allowed equal representation skew the audience (factor #2) who do not have enough background to discern the problem with that. We know that significant percentages of Americans believe that material widely considered to be pseudoscience is factual and that they get much of their information from media sources throughout their lives. So we regularly have the two factors at play. Yet, it remains unclear how the media's promotion of pseudoscientific ideas (as being on par with legitimate scientific ideas) influences the public and their views about the scientific process. Some studies exist regarding the effectiveness of pseudoscientific methods in marketing (Haard et al. 2004; Pitrelli et al. 2006). Survey research can also somewhat reliably tell us public opinion about how "scientific" they consider certain ideas. A regular survey done since 1979 by the National Science Foundation indicates American understanding of science and technology concepts and processes and also includes results about attitudes towards pseudoscientific concepts like astrology and creationism. 42% of Americans think astrology is sort of or very scientific[1] and 13% of biology teachers tell their students Intelligent Design is a valid scientific theory.[2]

It also remains unclear exactly how pseudoscience in the media influences the way the public perceives science in general. In other words, do people see ghost hunters on TV as doing respectable science? Do kids watch shows about monster hunters and hope to grow up to be a monster scientist? In the example of UFOs, author Daniel Thurs argued that public discussion about UFOs in the 1950s shaped valid science-related discourse in the public sphere. The topic was so popular that it prompted serious discussion about space travel and life in other planetary systems (Thurs 2007). In that case, a pseudoscientific subject may have had an unanticipated positive effect. Another example of a positive effect is the commitment many Bigfoot hunters have to wildlife and habitat conservation or that they learn how to collect DNA samples and have them analyzed.

The Pennsylvania Bigfoot Camping Adventure, May 2016. Brian D. Parsons is the speaker. Photograph by Kenny Biddle.

The Fringe

When scientists delve into inquiry outside the realm of the orthodox subject areas (sometimes referred to as "the fringe"), they run the risk of being ostracized by their scientific peers who see that as a useless, perhaps ridiculous, waste of time. Those that head into this unknown abyss may feel the risk is worth it because the payoff may be so large—a truly groundbreaking discovery such as ESP, or a giant unknown hominin living in populated areas. They may take the gamble and reject their peers.

Two loose labels for scientific study can categorized these fringe areas. "Deviant science" (Dolby 1979) is work in fringe areas rejected by orthodox scientists, such as ufology. Followers of deviant science accept the claims of self-appointed experts instead of considering the evidence and applying skepticism. "Anomalistics" (Truzzi 1998) is an interdisciplinary study of scientific anomalies or extraordinary events that do not fit with current orthodox theory, including pursuit of research in parapsychology, the paranormal, ufology, and cryptozoology.

Practitioners of both "deviant science" and "anomalistics" typically have scientific training and attempt to operate within the institutional rules of science. Therefore, the subject matter, not the process, is outside of conventional realms of research with one exception—the peers available for peer review are fewer, making the social aspect of this area of science insular. A closed community forms and evolves in isolation. Some researchers have examined deviant science from a sociological aspect (Dolby 1979; Goode 2000; Northcote 2007).

Why do certain scientists feel compelled to take this lesser-worn path? Complicated political, psychological, and cultural issues are at play. It is romantic, challenging, and exciting, and is good for getting attention (but not grant money). These aspects likely draw ARIGs to the same topics. People find it fascinating and the media gives these fringe subjects plenty of attention because of their oddness. We find scientists and everyday people given the title of "expert" within a small niche, becoming a big fish in a small pond, which has its benefits.

The Cranks

The dark corners of the fringe are inhabited by lone thinkers (almost always men) who are not just passionate but obsessed with one or more unusual and implausible ideas. The cranks. Anyone with an online presence who writes regularly about paranormal topics will have been approached by these self-styled pioneers and mavericks who proclaim they have made a breakthrough with an implausible idea about how things work (Collins 2006; Goode 2000). The cranks' tendency is to regard themselves as progressive or cutting-edge. And, they tell you so. Repeatedly. An exaggerated sense of self-importance about their work can be a hallmark quality of the pseudoscientist (Bunge 1984). Cranks can be self-educated, self-styled independent experts like Immanuel Velikovsky, previously discussed as the author who in 1950 interpreted ancient myths in terms of celestial cataclysms roundly rejected by scientists. Or, cranks can be PhDs that who have strayed away from the acceptable path, like John E. Mack (1929–2004), the Harvard psychiatrist who subscribed to and lent credibility to the idea of alien abduction of humans.[3] Cranks can clash with their academic institution over their work and its deviance from the orthodox mainstream. Mack was investigated, raising serious issues about academic freedom but ultimately prevailed.[4] In the case of Velikovsky, his work was published by a scientific textbook publisher, which outraged academics. The Velikovsky affair had ramifications still discussed to this day (Gordin 2012). Even negative attention paid to cranks will elevate them in the eyes of the public, giving them more readers and spreading their views.

New ideas that arrive from the fringes are rarely embraced by science (in what is called a "paradigm shift" where there is a wholly dramatic change in thinking), or the original wild idea may morph into something less wild and be subsumed into existing knowledge—consider the evolution of acceptance of heliocentrism, germ theory, plate tectonics, and quantum physics. But most often, ideas by self-styled modern Galileos are rejected due in no small part to cranks isolating themselves from reliable knowledge and critical thinking. Martin Gardner (1957) pointed out that cranks tend to create their own organizations, and self-publish their own journals and books—a substitution for the scientific community framework that rejects them—forming an isolated community defined by a specific worldview incompatible with mainstream scientific thinking. But individually or in fringe groups, truth cannot be judged by personal conviction. It must be supported by strong, unequivocal evidence, and a robust, predictive theory. The test of quality comes from the acceptance of conclusions by the scientific community and integration into knowledge over the long term.

Supernatural Creep and Conspiratorial Thinking

Naturalism, a scientific principle, states that natural laws and forces operate in the world, not supernatural forces. That is, explanations must be based on scientific laws, not invoked entities like demons or invisible spirits or forces that can not be tested or falsified (Pigliucci 2010). Use of religious or occult paraphernalia, divination devices, psychic powers, intuitives, spirit guides, or demonologists immediately signals that the investigators have moved outside the bounds of a naturalistic, scientific framework. While there are various reasons why these methods are utilized, a main reason seems to be to preserve and feed the belief. I call this tactic *supernatural creep*.[5] When normal processes and causes fail to satisfactorily explain strange events or answer questions, then reasoning slips by tiny steps to include excuses beyond nature, into super-nature, beyond the testable claims of science. It's a way to rescue a belief that the believer holds deeply. If you must choose between the belief or a rational explanation, the rational explanation may be that which gets rejected. When invoking supernatural reasoning at the same time as claiming you are using scientific methods, this is a clear warning that something is wrong with the methodology of investigation and is a signal of pseudoscience. This creep, which may also be viewed as a metaphorical "slippery slope," is apparent in all the ARIG subject areas. Bigfoot proponents exhibit this when straddling the line between natural and supernatural: there are those who search for a real animal that functions as nature intended and those who entertain the option that the entity is not natural (Gordon 2015). Those that embrace the supernatural explanation do it to explain why the creature has an uncanny ability to escape detection by eluding our cameras and leaving only tenuous, dubious traces of its existence. Perhaps it can run outrageously fast and may be able to see infrared light from night vision cameras or have other unusually developed senses. Some people resort to thinking of Bigfoot as a psychic phenomenon or related to UFO visitations because some witnesses state fantastic details such as the thing being immune to bullets, or suddenly disappearing. It's seen everywhere but found nowhere. How can we resolve this dilemma? The explanation for Bigfoot morphs beyond natural—it's not a flesh and blood animal, it's a shape-shifter, it can bend time, paralyze you, or disappear into a multidimensional portal. To explain the lack of fossil record for Bigfoot, a few even make the claim that the creatures derived from the Nephilim, the giant angels of the Bible.[6] Other cryptids have also been linked to occult explanations as demons, evil spirits, or mind manifestations. Cryptids are also linked to UFOs in several cases. UFO stories are commonly connected to supernatural concepts in their own right—time travel or distortion, dimensional travel, telepathic powers, and anomalous cognition.

After a long-standing emphasis on gadgetry and a sciencey focus, ghost investigations increasing conclude with religious or occult explanations—demons, angels, curses. One interpretation of this trend may be that science is failing them by not delivering the rock-solid explanation they desire. In order to maintain the belief, the natural explanations are no longer suitable. Parts of the cryptozoological community are moving down a similar path at the behest of authors like Nick Redfern who contend that the field should expand to include "zooform" phenomena. These are entities, not actual animals, that appear in animal form and have supernatural qualities. This would constitute a shift from scientific inquiry to a completely experienced-based view. Parapsychologists have even advocated changing the rules of science to move beyond naturalism to allow for looser explanations. This loosening is desired because, for decades, parapsychology has failed to be accepted under the stringent boundaries of scientific consensus. See Cardeña et al. (2015) for examples of this.

Certain locations may develop a reputation for having a plethora of spooky events including hauntings, poltergeist activities, sightings of strange creatures, UFO reports, and various other anomalous phenomena. ARIGs and paranormalists imbue these sites with special meaning, a supernatural aura, and describe them as areas of "high strangeness."

In a scientific sense, supernatural explanations aren't explanations at all (Pigliucci 2010). These ideas can't be supported; they invoke concepts outside the realm of natural law and thus outside scientific inquiry. One mystery is replaced by another. There is no net gain but the story becomes even more sensational, interesting, and marketable. As centuries of interest into these various mysteries yield no satisfying answers, fringe fields of cryptozoology, ufology and paranormal research continue to slide down the supernatural slippery slope.

A related corollary to supernatural creep is conspiratorial thinking. The answer to "Why haven't scientists taken cryptids seriously?" is thought by some to be "It's a cover-up, these weird animals are experiments gone wrong, the government doesn't want people to know." There are tales of the military doing experiments in psi operations or time-travel that resulted in catastrophes. The entire field of UFO research banks on the premise that the government is keeping things from us since 1947. The truth is some conspiracies are real, and there are some secrets that need to be kept for security purposes, leaving ARIGs without a means to get the information they feel they need.

Belief in intentional agents is one of the most common attributions for humans to make. Even the non-religious, who will not attribute events to God or the devil, seek meaning in patterns and may subscribe to conspiracy theories (orchestrated by higher-ups) instead of accepting random and complex explanations. Conspiracies and cover-ups are another way to provide a life preserver for the belief that can't stay afloat due to lack of reasonable evidence. Conspiracies are handy things in that they cannot be falsified—because you can propose anything to be true and no one can prove you wrong. It's all hidden; there is no way to confirm it. It's simultaneously the most ideal and the most worthless way of explaining a mystery.

Skepticism

Any discussion of the paranormal invokes the flip side of belief which is skepticism. Skepticism is roundly misunderstood and misapplied. "Skeptics" and "believers" are not exactly opposite ends of a spectrum as often portrayed. Some skeptics accept that there are indeed UFOs (unidentified objects) and understand that some people will interpret strange events as encounters with ghosts or unknown monsters. The difference is that skeptics will not readily accept a paranormal explanation. When the skeptic asks for evidence, and relies on reason and the tools of science, as appropriate, to examine the claim, this is seen as an adversarial act. To debunk or demystify, to pull away the mask, show the string holding up the UFO, and expose the hoaxers is seen as a bad thing. Questioning and demanding quality evidence are not inherently negative actions, though, and requiring sound support to make an informed decision is the basis of critical thinking. Debunkers provide a useful service to society by rooting out the truth from myths, misconceptions, and fakery. Skepticism—critical questioning and calling out nonsense—is one of Merton's scientific norms writ large.

ARIGs typically invoke skepticism in two ways—one positive, one negative. First, many ARIG members will say that they themselves are skeptical or was were once skeptical about the topic—the "avowal of prior skepticism" (Lamont 2007). Stating that they were once doubtful of this claim appears to strengthen their current position or conclusion by suggesting they considered all sides and were subsequently convinced (by experience, persuasion, or evidence). That's not necessarily true, though. It is more often used as a rhetorical attempt to establish the source as trustworthy and credible. This ploy also is a way to undermine alternative explanations and to persuade the listener to accept the conclusion provided. Often, it serves as a dramatic introduction to the explanation.

The second way ARIGs view skepticism is as a negative state. A skeptical person can be too doubting to the degree that she may create "negative energy" that could impede the investigation. This is one reason ARIGs aren't keen to have self-styled skeptics on an investigation. At least that's their explanation. One can surmise several other reasons why they regularly reject a skeptical presence.

Almost always missed in the ARIG concept of skepticism is the value of skeptical thinking as a method to find the best and most likely answer to a question, to eliminate bias and subjectivity as well as to apply logic and sound reasoning. Without skepticism in the practical sense, ARIGs easily fall prey to gullibility and wishful thinking that will make their work unsound. Without skepticism, you can easily be led astray from reality. Consider this statement: "[The] open-minded healthy skeptic considers that the paranormal explanation may be the more plausible answer."[7] The concept of open-mindedness is construed to mean a paranormal explanation should be considered a reasonable option. This way of thinking shows a profound misunderstanding of the natural world. Since we can never rule out *all* the potential natural explanations to reasonably conclude something is "paranormal," to call something "paranormal" is to give up on finding the real answer. It would be correct to say "I don't know," but the huge error ARIGs make is to take it a step beyond by saying, "I don't know, therefore, it's paranormal."

Skeptics who are willing to examine the evidence for a paranormal claim, will often ask for the *best* cases, whether that be for hauntings, UFO mass sightings, or cryptid encounters. In countless skeptical reviews of cases deemed to be genuinely paranormal by ARIGs and the public, critical analysis has revealed mistakes, flaws, and errors of all kinds. Skeptics will reject poorly supported and shabbily formed conclusions. Criticism is just an act of making a judgment (Nickell 2012) but the reaction to criticism is defensive and negative if it throws shade on the recipient's conclusion.

At a deeper level, Dewan (2013) says that the friction between skeptics and believers in the paranormal isn't so much about the reality of entities but about control over accepted views of reality. In this view, ARIGs have succeeded in making the paranormal view more acceptable to be discussed in public while skeptics remain as a voice of reason and a check to keep unsupported ideas from become too popular. Skepticism is most usefully thought of as a way to assess claims—a tool, an attitude, a method. ARIGs, however, may find it useful to use skepticism as a straw man or to couch skeptics as closed-minded to further their own positions. Eschewing sound skeptical practices makes your proposed conclusion far more likely to be seriously flawed and less likely to be taken seriously.

Conclusion: Beyond the Veil

Participation in an ARIG is a perfect opportunity to find "belief buddies"—one or more people around you that reinforces and strengthens each other's belief to further a shared goal or agenda. Dissent is discouraged in these types of belief environments (Pigliucci & Boudry 2013). But sometimes people make U-turns, spurred on by inherent curiosity, a genuine quest for the truth, a baloney detection filter, and, most helpfully, a friend of like mind who joins you in taking this new direction. What follows are two stories that illustrate belief buddies. The first is that of two former ghost hunters, Bobby and Jason, who took the path not taken by their colleagues, but not without resistance. The second is of Chris and Mark who stay on the path no matter what. The observations and actions of these two pairs illustrate many of the concepts in this book.

Once and Former Ghost Hunters

At age 16, Bobby Nelson began ghost hunting at his friends' houses. He had personal experiences in his own home that he interpreted, at the time, as paranormal. A devout Christian, Bobby was taught that if you believed in evolution, you were going to hell, that demons were real, and that there was life after death. Harry Price was his idol. William Roll was his mentor. He aspired to obtain a degree in parapsychology and started his own paranormal investigation group.

For Jason Korbus, it was curiosity about the unknown and macabre that drew him to paranormal investigation. From a non-religious background, he never had an experience that he would have labeled "paranormal" but was a fan of true-crime stories and TV shows like *Unsolved Mysteries* and *Sightings* that had actual scientists contributing. The stories looked legitimate, like a news broadcast. These shows reinforced his belief in ghosts. When he saw *Ghost Hunters* he was amazed people did this for a living. Why not give it a try? So he did and joined a paranormal group. Bobby and Jason related their time as ARIG members and paranormal believers in an interview with me in 2013.

In their twenties, Bobby and Jason became friends through their mutual interest in the paranormal. With their respective groups joined into one team, named "Phase 3 Paranormal," they visited people's houses in and around Toledo, Ohio, collected EVPs and electromagnetic readings, interviewed witnesses to the events and wrote up case reports, just like all other paranormal groups. They believed they had found paranormal activity on several occasions and that this was evidence of ghosts. EVPs, they concluded, were voices of the dead. Bobby recruited interested individuals for the group via Myspace, Craigslist, and the local paper. No special qualifications were needed to join although they administered an exam to new recruits to see how much they knew about the paranormal. Bobby remembers the ease in which people would give their Social Security number to him under the pretense of a "background check." This was his ploy to test for trustworthiness, since he had no means to run a legitimate background check. He figured if they would give him their personal information, they had nothing to hide.

Jason says they definitely were a "sciencey" group. They "absolutely" thought they were doing science—because of the equipment. For example, they would take a "baseline reading" of the house, Jason related to me, by walking around the rooms very slowly, waving the EMF meter around, and recording the numbers on paper. Later, after they "provoked" the ghost, they would do the same thing a second time and record the numbers. Bobby says: "For some reason, when you have that meter in your hand and you are looking for ghosts, that meter makes you feel like an expert. The piece of equipment in their hand that they think is giving them data they can use to somehow correlate with a ghost ... they feel sweet!" Equipment made them feel important and ARIGs judged each other by their equipment. While thermal imaging cameras (especially FLIR systems) were the epitome of the equipment bragging rights (TAPS, the *Ghost Hunters* TV show group, had a FLIR), the next best was the tri-field meter. "If you could afford a tri-field meter, you were badass," says Bobby. "I had a tri-field!" Jason adds. They would make fun of other groups who used dowsing rods or mediums as not being "scientific."

What was it about collecting EVPs, I asked, that was "scientific"?

"Sharon," Jason explains to me slowly in a fake patronizing tone, "it was a RECORDER and it was DIGITAL! We were getting voices from dead people!" At the time, they relied on such devices, it was no joke. They felt they were on the verge of finding something extraordinary, documenting evidence of ghosts. Believing that science was unjustly ignoring the paranormal, they were seeing it on display for themselves. While they didn't want to admit it then, they were basing their techniques and jargon on dialogue from the *Ghost Hunters* TV show. They assumed what was on TV was valid because everyone else was doing it too. They did the "reveal" for the client just like on TV. Jason remembers he used to always say, "We're here to help," exactly as on the show.

Half of their cases seemed to be attributable to haunted people rather than haunted houses. "I don't know what you found before, but this house is REALLY haunted," was a frequent repeat quote. Often they were told that other ghost groups had been to a location and reported "crazy" EVPs. Confirmed paranormal activity freaked out the residents and reinforced their fears. Investigators and the residents attributed everything that went wrong in the house to the "negative energy" present—health issues, bickering, even an abscessed tooth, was reasoned to have something to do with a paranormal presence.

Jason and Bobby mused about the paranormal mantra of "there are no real experts in this field" as "a bubble you can put over yourselves." It was an anything-goes atmosphere without standards. They couldn't help but wonder how people on different teams could get such different results. Why didn't Bobby and Jason see ghosts like others did? When they began asking questions, the façade began to erode. They couldn't find definitive answers regarding a paranormal cause for electromagnetic field readings. Other explanations seemed more plausible and tested positively. When they did flashlight tests and EMF readings at non-haunted locations, they got similar results as in a reportedly haunted location. Jason described a simple test they set up to check best-quality, supposedly unmistakable "Class A" EVPs: "We'd play an EVP for five or six different people and we'd say, 'Write down your answer independently. Don't tell us what you think its saying, we'll go over it later.' We'd get six different answers. Remember these were supposed to be Class A unmistakable EVPs...."

Bobby owned a recording studio. He took some EVPs to his sound engineer who was only able to tell him that it was within the range of human hearing but not if it was anything unique. He certainly didn't say they were paranormal. They recalled one incident where Jason knew he'd zipped up his backpack during an investigation. On the audio playback, another ARIG member reported they captured an EVP that sounded like a voice saying, "Del Rio." Maybe these EVPs were not all that reliable. If it was this easy to mistake a real sound for a ghost voice, what about all the other "evidence" they had?

Their new inklings of doubt made the two uncomfortable. Bobby called them "woo woo moments": "I would say 'Am I being too skeptical here? Am I being so skeptical that I'm preventing my brain from seeing paranormal activity?'" There was no final "Aha!" moment for either of them; it eroded away little by little, gradually slipping away.

On October 31, 2008, *Ghost Hunters* aired a show live from Fort Delaware. The episode prompted an outcry from viewers that some of the evidence had been faked. The team that had inspired thousands of others to do paranormal investigation was accused of duping their audience, fans, and followers. This event was further material for Bobby and Jason's evolving views. "We were disillusioned specifically with the team that we looked to for inspiration," Jason says. "Maybe all this stuff on TV is fake."

Bobby says Jason "gave up the ghost" before he did. When Jason admitted he didn't believe in it anymore, Bobby was upset. "I was fucking crushed! It killed me inside to hear him say he didn't believe in ghosts." Yet Bobby was also well along the path of skeptical thinking.

Peer influence and community interaction affects how we relate to issues in our society. In the case of Bobby and Jason's paranormal investigations, they had been influenced by pop culture and the paranormal community. Now, their circles of influence were changing.

Bobby is proud of the regular phone conversations he used to have with William Roll, an esteemed parapsychologist who investigated poltergeists and haunting cases. Roll died in January of 2012. Roll had mentioned James Randi in his conversations in a not-very-complimentary way. Bobby wondered, who was this guy, Randi? He was, in fact, one of the world's foremost magicians, skeptics, and arch-nemesis of pseudoscientists. Randi agreed to come on Bobby and Jason's radio show for an interview. Bobby recalls how he tried to nail Randi with the standard sciencey tropes such as invoking the law of thermodynamics. It's an embarrassing memory to them now, as are many of their past public pro-paranormal pronouncements. Jason and Bobby consider Randi to have been a critical influence on their change in thinking as well as Michael Shermer, Ben Radford, and Kenny Biddle (another ghost-hunter-turned-skeptical-advocate). The tone of their radio show/podcast, *Strange Frequencies Radio*, has changed drastically over the years as they transitioned from paranormal believers to paranormal skeptics. The supernatural and paranormal ideas had all evaporated. Their enthusiasm and curiosity, however, had not.

There was clearly something about their friendship that played a part in their individual journeys from paranormal advocates to counter-advocates. I asked them how much of an influence they had on each other. They both agreed it had been significant. They had reinforced each other in the practice of questioning, examining, and gaining new perspective just as they had previously reinforced each other's belief in the paranormal. Bobby would ask questions, and then would buy a book about it. Jason would borrow the book. They would discuss their new ideas. "It was good to have Bobby there—the only one willing to go down that road with me," Jason states, "Anyone else would get hostile." No criticism was allowed in the ghost hunting clubhouse. It still isn't. If no critique is allowed, no mistakes are ever corrected. Thus, no progress is ever made. And that's how it currently stands, years later, with popular paranormal investigation.

The case reports from "Phase 3 Paranormal" investigations remain in binders and in boxes. Media requests dried up when their new stance became "demons don't exist" and "Ouija boards aren't a portal to the afterlife."

It's easy to believe. A much greater effort is needed to be skeptical and to examine all options for the best answer. Jason and Bobby had to stop and re-train themselves to think in this new way and let go of a previously sacred idea. Bobby discovered and accepted the explanation that his paranormal experiences long ago were not demons but episodes of sleep paralysis which he still occasionally experiences. He says he was comforted, not disappointed, by the reality-based explanation. Many people invested in paranormal belief and research will not be able to let their decades of emotional investment go. Bobby reasons that he didn't have as strong of an emotional tie to paranormal ideas as some people. He understands that people don't want to accept their time and money has been wasted. And to some, there is a deep-seated need to validate the afterlife—"that orb is Grandma." No matter what.

Jason is clear about his beliefs: no ghosts, no paranormal, no supernatural at all. "I don't care that people believe that stuff but keep your hobby out of other people's houses." Both men regret that they may have done harm to people by telling them what they thought was true at the time. They thought they were helping. Now they hope they are helping spread critical thinking about the paranormal by sharing the evolution of their own philosophies. "Don't visit your BS beliefs on other people," Jason warns. "Don't make the same mistakes as us."[1]

Into the Swamp

In July of 2016, I was interviewed by a journalist who was profiling two men seeking proof of the skunk ape, Bigfoot's smaller cousin, in Florida. The writer, Bill Kearney, talked to me before he ventured into the Green Swamp with the monster hunters whose interest in the cryptid was piqued by the show *Finding Bigfoot*. According to my model of ARIGs that I'd developed by this time, I had a good idea of what he would find. The resulting piece indeed revealed that these guys also fit the typical ARIG characteristics.[2] Chris and Mark admit they struck out on their quest as "a big hobby" and "therapy"—something interesting to do to get away from uncomfortable realities. They gave creative names to portions of the swamp, "Creepy Hollow" and "Thunderdome," where they believe the skunk ape is living. They say it's watching them—they hear it, they feel it. They "know" what it eats, how it behaves, and what it's thinking. Via their videos posted to their YouTube channel, they collected a small following of fellow cryptozoologists. The men believe that skepticism expressed about this topic has scared off scientists and they decry skeptics (particularly me) who have never been out in the swamp to see for themselves.[3]

"We're not into the paranormal crap," they say. Yet, their descriptions are not of normal natural events. "We hear voices, stuff that says our name. They hear us talk. It's mimicry." Mark says he saw a shimmering figure like from the movie *Predator* in the swamp as well. "I didn't see it again, but I know what I saw. I definitely know what I saw."

Both Chris and Mark think humans are arrogant in our knowledge. They relish the mystery and admit they maybe don't want the skunk ape (or whatever is out there) to be discovered.

Chris and Mark are not unique. They are two of thousands of seekers out there looking to find personal meaning within this effort. It's not about facts that you or I can check and agree upon. When emotions and strong beliefs rule the heart and mind, a scientific process doesn't work. There is nothing I can say to Chris and Mark that would sway their worldview that relies on the belief that the Green Swamp is genuinely mysterious and contains a skunk ape or two.

Dr. Bryan Sykes concluded after spending time with cryptid enthusiasts who really believe the truth is out there, that one can get swept away in the dream that something big is just out of reach, ready to be revealed (Sykes 2016).[4] People can make the turn away from nonsense and towards a more solid worldview built on evidence instead of TV pseudo-reality. The story of Bobby and Jason illustrates this. It certainly doesn't mean giving up your interests, but it does require revising your worldview.

Inside-Outside

My study of ARIGs across the Internet and in real life revealed a worldview very different from my own, but shared by a community of thousands and many more thousands around the world.

Paranormal beliefs are embraced by many and dismissed as nonsense by others. Paranormal believers' worldview is that of an alternative reality—one that includes other senses and dimensions besides the ones we know and rely upon. It is unrestrained by borders of facts, proof, science, or reason. Loxton and Prothero (2013) discuss this in their last chapter, "Why Do People Believe in Monsters?" (The same reasons apply to ghosts and UFOs and other paranormal concepts.) People believe in what seems unbelievable because it provides value to them regardless of the possibility of being seen as unsuitable. Those that embrace the paranormal can achieve a sense of control and importance over their lives and feel less threatened by the chaos, fear, and sadness. This alternative worldview provides them a meaningful benefit.

Even the unbeliever can come to adopt these ideas via small steps and eventually hold a supernatural worldview. Belief in one fringe idea leads to exposure and acceptance of other similar beliefs. Fortunately, the same small steps can also go the opposite direction towards critical evaluation and acceptance of science and reason.

One might adopt cultural beliefs in order to belong to a particular social group and experience a thrill. If the paranormal belief is enforced by a group or tribe that the individual identifies with, it will be exceedingly difficult to give up that belief.

Bader, Mencken and Baker (2010) make clear that it's a mistake to cite sweeping generalizations to describe the current paranormal culture and the people who subscribe to it. There are complex sociological and psychological reasons why a person will accept a paranormal explanation. Television, movies and the Internet have made gigantic impacts in our culture. Portrayal of paranormal belief is mainstream and generally acceptable. People now have easy access to learn about a subject, plan for experiences, and become involve with others that encourage belief.

Complex stories don't make for good media content (Bader et al. 2010: 75). Therefore, the audience misses the nuance, the backstory, and the necessary foundation and support needed for full understanding of research, events, and conclusions. It is easy in our society to claim the role of expert when expertise is fundamentally lacking, to be a teacher when experience is shallow, and to portray a scientist without substantial education in any sciences. This social loophole allows for deception and possible harm; when critical thinking and earned expertise is not assigned high value, actors and pretenders will lead us by the nose.

My goal in counting and describing ARIGs was meant to illustrate how passionate, dedicated, determined, and creative they are. I needed to show that ARIG participants are not marginal or stupid. They reflect our own culture that values authoritativeness, scientificity, and a sense of wonder and mystery. The paranormal community has embraced individualism as well as finding stability and usefulness in shared ideas. They have created their own sub-communities, argot, and special knowledge. They have their own sacred places to visit and individuals to venerate. The influx of participants throughout the 2000s signaled a collective desire to find meaning and have experiences beyond everyday life. Performance and participation became essential components of ARIGs (A. Hill 2013) With ARIGs, paranormal culture became domesticated and democratized. It's everywhere and for everyone (Jenzen & Munt 2013). It still remains a source of mystery and general interest for the public and is a means by which some people create meaning in their lives and undergo personal transformation (A. Hill 2010). People who participate in group activities or TV shows create their own narrative in what might be analogous to live-action role playing (LARPing), transcending their day-to-day persona and boosting their self-esteem and sense of worth. They see themselves as "monster hunters," "ghost busters," "afterlife warriors," and "truth seekers," eager for an authentic experience to "see for themselves" (A. Hill 2010).

Throughout the history of investigation into these topics, scientific interest has waxed and waned. Influential people and ideas have come and gone. Some threads continue to weave through and remain relevant but most fade away. The newcomers wish to jump right in and neglect valuable historical information. The public, always fascinated by the unknown and mysterious, discovered that ARIGs filled the empty void scientists left when they abandoned serious interest in the paranormal. Those who made it their serious leisure activity gladly assumed the role of amateur expert. Mayer (2013) studied the "phenomenology" aspect of American (and German) ghost hunting. He noted that the paranormal investigators are socially situated in such a way that they are both "insiders" and "outsiders." They advocate use of "science" but are untrained amateurs. They claim to be skeptics but promote belief. They desire fame and celebrity but brand themselves as the everyman just asking questions. They say they are seeking truth while promoting unsupported, implausible ideas. And, they dream of financial gain while boasting that they do not charge for their work.

Short-circuiting of the formal process for becoming credentialed as a teacher or expert is detrimental to the public. One could even argue that such activities promote scientific illiteracy as people mistake what is being explained as corresponding to genuine scientific terms and methods. The audience is delivered inaccurate information, a distorted view of science, and an unsound investigational process devoid of a connection to history, existing scholarship, and cultural context.

Science is...

A discussion about the correct nature of modern science was necessary for this examination of ARIGs to compare what they do as "science" to an accepted scientific methodology.

The concept of science as method, process, community, and body of knowledge is not always grasped by the lay public who see science more as a certain kind of academic person or what's in a textbook. Science requires using the most reliable methods of inquiry to obtain reliable knowledge. This multilevel, nuanced understanding of the scientific endeavor is lost on ARIGs just as it is invisible to those who never pause to examine the foundations and norms of scientific research. The average person has no knowledge of how science really works because we are never taught about it—a major oversight in the education system and one that hampers scientific literacy across the world. Today's popularity of paranormal investigation is not, however, just a failure of science education or lack of teaching critical thinking skills, it has much to do with psychosocial and sociological factors. The portrayal of science by ARIGs also provides us with important insights into the public perception of what science is and how it works. Science as a cultural concept is different than science as applied by active scientists. Science is a construct as well as a real, complex endeavor in human society. It also has a strong social, even political character (Dolby 1975). There are only so many social resources to go around—money, attention, authority, respect. Scientists get defensive if a charlatan or faker attempts to usurp those areas that historically have been the realm of scientific endeavors and tries to undermine the authority of science. However, a great deal of work is required to do research up to scientific standards. Most ARIGs do not have the resources, preparation, or opportunities to push their work to a scientifically acceptable level. Since it takes a significant investment of time, money, and perseverance to attain a higher degree or coveted letters after one's name, scientists are keen to retain their rights and ownership of the authority of science and resent those who attempt to fool the public into thinking their abilities and pronouncements are of equal merit. But the public displays a desire for science to be more democratic, for the average person to participate and have a say in what is explored and investigated.

I observed how ARIGs retreated from their strong public "scientific" stance and quickly qualified their processes when confronted. They did not rise to the challenge to present, support, and defend their evidence. Instead, they often resorted to the weak and meaningless counter for the skeptic to "prove it *doesn't* exist"—we could go round and round forever and get perfectly nowhere with that end goal.

ARIGs lack communalism. Instead, we have hundreds of different interpretations of data and sets of conclusions, little to no cooperative efforts, and seriously flawed data sets interpreted by researchers with an entrenched bias for a particular outcome. Continuing to communicate only in their small groups who subscribe to the same beliefs and methods, ignoring criticism and requests for clarifications, ARIGs risk painting themselves into a smaller and smaller corner with no satisfactory way out except to produce even more unacceptable excuses. Their hypotheses and theories remain undeveloped and poorly tested, if tested at all. The conclusions serve as creative but unsupported explanations that sound interesting but are more fiction than fact. Many seemed to be explicitly seeking confirmation of their prior belief in ghosts, hauntings, cryptids, or UFOs as paranormal.

ARIGs have taken advantage of several social aspects relating to science to gain advantages with the public. By using sciencey appurtenances and appealing to democratic sentiments about knowledge, they can effectively serve in the role of scientist in their specific subject areas. Laypersons have a difficult time judging expertise, especially in these times when incredible volumes of information are available for little to no cost. The public is free to choose their "expert" based on their belief preferences. As Francis Bacon, a founding father of science, said: "Man prefers to believe what he prefers to be true." Someone who sounds suitably impressive will stand up for any position when it serves some purpose. Scientists have been found to argue against clear-cut theories such as evolution, climate change, vaccination efficacy, cigarette health effects and many fringe paranormal ideas. Credentials, especially diplomas and celebrity status, become highly effective tools to garner trust and followers. Since scientific concepts are difficult to comprehend and evaluate by those not educated in such fields, laypeople rely on trustworthy sources for acceptable information. And they deem a source trustworthy based on clues and cues. Even small portrayals of scientificity are influential in accruing authority and trust from the lay public. When a preferred paranormal belief is supported by an apparently authoritative person who sounds sciencey, people feel justified in greater acceptance of it and they use the sciencey aspect as a basis to convince others of its legitimacy (Blancke et al. 2016). This results in pseudoscience being given equal weight to scientific concepts in the public sphere.

Blancke et al. explain the problem at the center of belief in these topics and why being *scientifical* is so effective as a strategy:

> People are not interested in impartial truth, but in finding and supporting beliefs that make intuitive sense.
> Pseudoscientific concepts are pervasive (1) because posing as science works as a

tool of persuasion, and (2) because people lack the motivation to correct their intuitive beliefs, but instead seek to confirm them and, simultaneously, distrust genuine scientific expertise. Various factors explain why cultural mimicry of science is a successful strategy for irrational beliefs to adopt: the fact that science is a dominant cultural authority (mimicking science can only be a successful strategy in a culture that already holds science in high regard), the exploitation of epistemic vigilance, people's misunderstanding of science's authority, the use of science as an argument, and epistemic negligence.

With working scientists untrained and unwilling to do public outreach to educate or to correct misunderstanding, society is faced with the conundrum that Carl Sagan described in 1990: "We live in a society exquisitely dependent on science and technology, in which hardly anyone knows anything about science and technology. This is a clear prescription for disaster."[5] As much as the public would like to participate and support scientific conclusions they prefer, science is not a democracy. An idea promoted by a few does not have equal weight as those theories that have been well-tested and are strongly cross-supported by tenets and facts from other areas. Science is also not like a trial in a court of law. The analogy to legal arguments is commonly made by ARIGs who say that their evidence would be acceptable in court to convict a person of a crime. Science relies on more than just a preponderance of evidence. To establish natural facts requires consideration of existing knowledge and achieving a consistency of explanations. Fields like ufology, ghostology, cryptozoology have remained marginal because of the less-than-rigorous methods that fall short of scientific standards.

Science Appreciation

Loxton & Prothero (2013) individually presented their view on cryptozoology in relation to science and education. Daniel Loxton feels that the investigation of Bigfoot and cryptids is a "gateway" to scientific literacy. As a children's writer and illustrator, he has observed that the monsters are interesting and can kick-start a conversation with kids about biology, evolution, and evidence. On the other hand, Don Prothero, a prolific author and professor of natural sciences, sees cryptozoology ideas as promoting scientific misunderstanding. It's bad enough that science illiteracy is rampant in the U.S., silly ideas about a real live ape-man roaming our country undetected without leaving a certifiable physical trace impede the reality of accepting scientific truths about nature. Considering my exchanges with ARIGs, I agree with both views but they apply to different audiences. Critical thinking skills can be learned and applied by everyone of all ages, in all stages of life, and to many important activities in life. It is essential to start teaching these skills very early, such as in elementary education. Use of creative topics such as hauntings, monsters, and UFOs to illustrate how to think through these claims critically is a winning approach. Kids love it, they relate to it and they remember it, planting in their minds a pivotal idea that what the public thinks is true may, in fact, be more complicated or outright false. It may instill a desire in some students to probe claims with further questions and nurture skeptical tendencies. Once the mode of thinking about claims is set and the person is beyond formal education, it will be more difficult to instill new practical skills. Therefore, constant media attention to these paranormal claims as valid, placing pseudoscientific ideas in equal comparison to scientific conclusions, reinforces mistaken ideas as true.

I see high hurdles to making progress with public appreciation of science, its methods, and conclusions. The notion that science is outside the average person's purview is detrimental to science appreciation. Society must be comfortable with the scientific process, trusting that it is the best method we have but accepting that conclusions are always subject to revision. This is an incredibly difficult sentiment to accommodate. Most people want an answer and are unsatisfied with probabilities.

I found many examples with ARIGs that reflected a confused view about science in relation to definitions and norms accepted by the modern scientific community. Confusion was especially obvious when occult or religious beliefs are mixed with "scientific" protocol. While it isn't a problem for a scientist or investigator to hold religious or occult beliefs, when those beliefs influence the collection of information and conclusions made, then the tenet of scientific naturalism has been breached. These groups are seeking success on their own terms which fall short of scientific endeavors. As a society, we can improve and learn to think well in general. This would also benefit paranormal investigations. If we could decouple the idea of the idealized "scientific method" from the larger concepts of science and critical thinking, I believe more progress could be made. ARIGs can make sound decisions without scientific training or pretending to be scientists. This has been shown to be true by the activities of the few skeptical-minded ARIGs. It will not be possible to inject science or critical thinking into the discussion with those who seek enlightenment via the paranormal. For those ARIGs that have a spiritual goal, science is not welcome.

Sham Inquiry

ARIGs see bits of evidence differently than others would as they employ a particular framework to interpret it. Pieces of evidence should all come together independently to tell a coherent story that does not fly in the face of what we already know to be true. The evidence should form an explanatory picture with the pieces mutually reinforcing each other (Haack 2003). The paranormal worldview results in evidence that appears to fit a framework that reasonably supports the preferred paranormal cause. This framework provides a pre-existing shape like a wire frame. It skews the investigation and conclusions resulting in a non-scientific and disputable result. The process of inquiry that is undertaken (perhaps unconsciously) results in gathering only "evidence" that can be fit to this particular frame or preferred narrative. Non-conforming evidence is ignored or explained away. The script has already been written, the destination set, so the process of "investigation" is fail-proof. It's sham inquiry. The sham inquirer knows just the right kind of evidence he seeks. Seek, and he finds.

If we put weak or misleading evidence in, we must consider what comes out. Garbage in, garbage out. Volume of evidence matters far less than quality. A few carefully done studies are preferred to dozens of low-quality studies that may all be wrong. My favorite analogy for paranormal evidence was made by Benjamin Radford who said that you can't make a strong cup of coffee by pouring together many weak cups.[6] And you can't make a strong case with a lot of weak evidence. Context is also important: observations or recordings out of context are worse than useless, they are misleading.

Carl Sagan, writing on UFOs, recognized the lure of fringe ideas. They are charming, would be delightful if true, and carry deep emotional significance to us and our place in nature. Like Harry Houdini who exposed psychics or James Randi who exposed faith healers, Sagan was aware of how strong emotion can overcome our senses and allow us to be deceived. We believe weak evidence simply because we want it to be true or we reject it when we don't want it to be true. Sagan made an effort to pause and examine before believing or rejecting. We must always strive to do the same (Sagan 1972). His maxim that extraordinary claims require extraordinary evidence still holds.

ARIGs collect a lot of weak evidence and it may also be pulled from and applied outside a reasonable context (in a paranormal framework). The evidence goes unscrutinized, and is scientifically wasted, lacking coherence with any reasonable overarching model of what could be going on. A paranormal interpretation goes against the rest of reliable knowledge we have, making it highly implausible to be true. The information may be narrowly presented to appear to converge to a paranormal entity, but in the bigger picture of nature, it does not. Ideally, all the evidence ARIGs produce should be critically evaluated on merit with poor stuff tossed and the remainder used to form an independent hypothesis and to generate testable or answerable questions. If a large animal exists here, then I should find certain things in the environment that correspond to that conclusion.

If this is a haunting, then I should find consistent results between investigators at the same place. ARIGs do not formulate testable hypotheses very often because it's not their primary goal, which is, instead, to prove the paranormal. Therefore, this work isn't sound research.

Most ARIGs are not at all aware they are taking this route of sham inquiry. Instead, they assume they follow a systematic approach that will get them to a solid conclusion. They often demonstrated that they believe they are using the best science and reason they can. The process feels satisfying to them and they get rewarded with their expected result. Going through the motions but not the intellectual effort of inquiry scuttles any chance to get a robust result. When they find the evidence that fits their anticipated result, they stop, and inquiry stops too. At that point, acceptance of a preconceived, probably paranormal, explanation is assured and hard to budge. Prior commitment to a specific belief is a block to even considering other evidence that may conflict with that belief.

The ARIG will work to maintain the conclusion in the face of critical questioning. They are not open to having that belief challenged and will typically be hostile to skeptical processes or persons whose role it is to do that. In this behavior, ARIGs exhibit tribalism—adherence to the values and practices of their community. Scientists and other communities all fall into tribalist tendencies (McRae 2012). But tribalism can get out of hand, causing the tribe to lose touch with objectivity, ignore important elements, and become hypocritical.

This behavior manifests in social media interactions among ARIGs and between them and skeptical commentators. Tribalism is another hurdle in moving towards shared knowledge.

Even though it feels satisfying to make a conclusion that this or that anomaly is attributable to a paranormal cause, that's not a valid investigation. If failure to meet a reasonable standard of investigation is deliberate or unintentional, it is intellectually dishonest if it's presented as a fair inquiry into the issue. After decades of research by amateur investigators who seek ghosts, UFOs, and cryptids, we observe that they have not improved their methods to the point where better evidence is obtained. But that doesn't mean good research *can't* be done by ARIGs. A scientific background is not necessary to increase the quality of such investigations. Acceptance of some basic guidelines and goals of critical thinking is required for ARIGs and interested individuals to make progress and contribute to understanding seemingly paranormal experiences.

Rational Paranormal Investigation: The Alternative to Sham Inquiry

A great deal more effort and time is required to do rigorous, meaningful research and investigation into ARIG subject areas. The goal of improving research methods and conclusions may not be suitable for many ARIGs who prefer to continue to do what they do now. There is no real incentive to do anything different, as debunkers are not rewarded nearly to the degree that the paranormal view is. The media, seeking drama and controversy, latches on to false positives. ARIG clients may wish to validate their belief instead of gain a reasonable explanation. ARIG participants may be drawing rewards from his or her serious leisure activities and belief in paranormal ideas. Popular culture will elevate the "better story" for its needs. A cultural shift towards heightened value for critical thinking is required before we see due respect for more thoughtful investigation and credible expertise.

If ARIGs do wish to tighten their methods and limit biases that prevent sound research results, if they wish to do rational inquiry instead of sham inquiry, the following sections are a guide to a revised approach. If information is collected carefully and then shared, considering the core scientific norms, amateur non-scientists can contribute significantly to the body of reliable knowledge. As with citizen science projects, evidence and data can be presented to credentialed scientists who use it to move along the path to produce a scientific conclusion. Scientists better serve the public with regards to these processes (Sykes 2016) since they can provide the time, funds, and expertise to conduct large studies and statistical evaluations—things ARIGs can not do but that are necessary for scientific rigor.

Loxton and Prothero's *Abominable Science* (2013) provides some goals for cryptozoologists to improve on their research. The tenets apply to all paranormal investigation and are as follows:

Rethink fundamental assumptions. As described thoroughly in this volume, ARIGs often come into an investigation with preconceived notions and strong beliefs in an entity. This short-circuits the investigation process immediately.

Test the null hypothesis. Is there enough evidence to overthrow conventional explanations in exchange for the alternative? Is there anything to explain?

Meet the burden of proof. Strong, positive evidence is required when making a claim that could more likely be a hoax or a misinterpretation.

Collect high-quality data. Credible physical evidence and logical analysis is needed. More stories and eyewitness tales aren't going to cut it (the weak coffee analogy).

Publish high-quality reports and work. Proper documentation is needed for the wider public to access and judge the claims and conclusions and to build on that work.

Be open to criticism. Peer review is a necessary component of science. It is a harsh process, no one likes to be criticized, but it must be undertaken to gain credibility and fix errors. Ignoring skeptical literature and reasonable criticism reflects poorly on fields that strive to be taken seriously.

Be skeptical of your own data. ARIGs do not recognize how easily they can be fooled and have regularly been duped by those with alternative agendas. They must learn to harshly judge the quality of their own evidence and concede alternatives and flaws. In the end, the ideas must form a solid theory or model that explains and predicts. Facts must fit into that model and the resulting explanation needs to made make sense without requiring baseless assumptions. Research must fill in the gaps and strengthen knowledge, not aim to upset the entire table.

Finally, we all would benefit from *accepting uncertainty.* To invent answers is to create misinformation. It's really OK to say "I don't know" but let's continue to look for the best answer.

Reassessing Goals

Generalization as to what ARIGs discover is difficult because each person, each group, and each community is seeking their own understanding as it applies to them. Sunk costs (the time and money invested into a cause) make it difficult to walk away from the search. An alternative to doing things the way they have been done for a decade or more is to reassess the goals, and reboot or retool the effort in a way that makes the work still personally fulfilling for ARIG participants but provides useful knowledge to the rest of society.

Careful research, record keeping, thorough documentation, and archiving should be goals of ARIGs who say they are trying to contribute to science. A reliable, high-quality effort in cataloging reports of paranormal claims would provide a means to classify a phenomenon, spot patterns and similarities, and perhaps formulate predictions that can be tested. Such a collection would be valuable for social and historical context alone.

Just about anyone can be trained to think more critically and carefully. I offer to consult ARIGs gratis to help them sharpen up their science understanding and improve their methods. I have had two participants in this program. My method involves a first step to establish clear goals and expectations, and plan a path towards achieving the goals. I provide them a series of questions to answer in a group discussion. This first step is daunting and they invariably give up prior to fulfilling it. I suspect they reject the step because such personal reflection is difficult and they realize that doing so would mean going down an uncomfortable path, so they say "forget it" and continue with their current ways. But, I contend that any successful project requires preparation, otherwise, the effort is just a lark and it will not work out well. For ARIGs to honestly assess their reasons for doing this and recognize their inherent biases, an outside motivator is needed or internally-motivated reasoning will quickly take the group off this tough course. The following discussion questions are provided to ARIGs and they are asked to write the answers or otherwise record their discussions for me.

1. What is your goal on each investigation? (Be very basic, short and to the point.)
2. Do you consider yourselves scientific in your approach? What does that specifically mean to you? Or if not, what approach do you utilize—spiritual, metaphysical, etc.?
3. Describe your typical group members or people who are most involved with investigations. Do any have scientific training?
4. What is your understanding of each piece of equipment you use (including personal feelings you record during each investigation)? What purpose do they serve and what is your basis for using it?
5. Describe your best evidence and why you think it is convincing. Should I be convinced? Why?

Framework for Paranormal Investigation

The five questions in the previous section are meant to make the ideals and values of the ARIG transparent so all participants understand. To break from the parade of cookie-cutter investigators who unfortunately don't produce any research and conclusion of substance, methods and attitudes must change. I provide a new framework for ARIGs to use that is superior to the TV-based sham inquiry method. As part of this process, I recommend ARIGs study the skeptical literature. (For a good start, I recommend Benjamin Radford's *Scientific Paranormal Investigation* (2010) to contrast the characteristics of a solid investigation with a made-for-TV one.)

Conclusion

It is imperative that any investigator understand these basic concepts:

- Chance, randomness, and correlation vs. causation
- Characteristics of scientific research
- How we fool ourselves (bias, perception errors, and unreliability of memory and anecdotes)
- Critical thinking processes

Establishing a clear goal for each event (investigation, ghost hunting trip) will influence the way the investigation plays out and whether it is ultimately satisfying, frustrating, or a waste of time. (Is the goal to have fun? To enhance your belief? To find out if a claim is true? Each of these will require a different approach.) The goal also will illuminate what constitutes a useful result.

Think about the practical utility of this investigation and whom it benefits. Is the aim to get to the best answer about a claim? Will others who were not part of the investigation find the conclusion satisfactory and useful? What would be the objections from critics?

Establish the claim clearly and narrow down the questions you are asking about it. Make sure there is something described that you can examine for an explanation. Seek to illuminate the issue with verifiable facts and explanatory details. Stick closely to this claim for investigation instead of looking for ways to enhance it.

Decide if you are using the right tools. What's the intended use of each tool? Does this match the use for which it was designed? If not, what is your justification for using it in another way? Can you explain what data it gives you and what that means? Find references for these answers. If they are paranormal-themed references only, find objective technical references for the devices or claims about how they work. If you can't explain and defend why you are using a tool so that others outside of the paranormal field will understand and except it, ditch it and go back to basic observations, note-taking, and fact-gathering. These exercises in thinking through what you plan to do and why are key to effective investigation, but hardly any ARIGs seem to do this.

With regards to process of an investigation, the following considerations should be emphasized:

- Work on establishing the claim exactly through interviews, research, observations.
- Do proper scholarship. Get facts and quotes right. Seek primary sources. Corroborate facts with evidence. Trust but verify.
- Utilize critical thinking, evidence analysis, logic, and established

scientific methodologies. Identify assumptions that, if false, make the mystery disappear. (The witness may be hoaxing/lying, or they may make a mistake in facts or observation.)
- Look for negative evidence. What is missing that should be there? (Did the dog not bark? Are there no body or physical traces of a reported animal?)
- Eliminate natural explanations as best you can. Every place will have some unexplained activity because we are missing information. It is impossible know every possible explanation. Therefore, the logical solution to an unexplained case is "I don't know." Use this when necessary.
- Consider ethical concerns. Never go beyond your limits (people with psychological or health issues, children). This is harmful, morally questionable and legally treacherous. It's unethical to suggest causes that are implausible and promote your personal belief (demons, angels) and to spread misinformation or fear.
- Do not assume the existence or behavior of something without evidence to back it up. Get independent confirmation of facts, suggest more than one possible hypothesis and examine each. Beware of lumping observations together, or correlating them to a single cause (haunting). Don't rely on any explanation by default (such as "I can't explain it, therefore it's paranormal.").

Avoid these mistakes:

- Assuming all methods are valid. Even though other groups may use a technique, it may be nonsense. Some equipment and processes are bogus, leading to useless evidence.
- Considering "feelings" as evidence. A person is a subjective "tool." Our senses are frequently unreliable. Recognize that contagion may be at play—if you say you feel cold or ill, others may be sensitive to suggestion.
- Failing to consider alternatives adequately. Look for natural explanations. Also, consider: Why does it have to be a ghost? Can it just as likely be aliens or humans from another dimension?
- Avoid unproven paranormal theories and hypotheses. No one person can reasonably conclude something is "supernatural" because comprehensive testing and systematic reduction of all possibilities is untenable.
- Counting on orbs, EVPs and any video or audio as best evidence. There are innumerable other possibilities for explanation including equipment glitches.

Write up your results completely:

- A complete record of all details is required.
- Include all carefully recorded notes.
- Avoid making assumptions, just rely on the facts.
- Analyze the findings, doing follow-up research if necessary.
- Use references and cite them.
- Make the report available to discuss with others. Request a review and critique. To make the strongest report, seek out a skeptical reviewer.

Innovation and Growth

Most ARIGs are closed to opportunities for critique, cooperation, and change. Engage with those with different views instead of being hostile towards them. A variety of viewpoints can only strengthen your knowledge and help build a more solid conclusion.

A well-run ARIG requires a regular evaluation of the group mission, attitude, and mindset. Avoid copying other groups and expand knowledge independently. Investigative techniques should be promoted that break down the problem, seek high-quality evidence instead of anomaly hunting, and reach a sound conclusion. The desire to support a pre-conceived idea about a paranormal subject must be discarded instead of encouraged.

Societies and organizations are vital to producing a set of standards for researchers to follow and training in those standards. ARIGs certainly benefit from having a community and should make efforts to share data. An overarching organization that sets recommendations for methods, encourages peer review, and facilitates publication of results would be highly beneficial.

Change is uncomfortable. Stagnation is a waste of time. Bad methods are useless or harmful. ARIGs can change for the better by eschewing "being scientifical" and by recognizing how to avoid sham inquiry. The alternative process requires deep reflection, additional understanding, diligent work, critical thinking, and openness to criticism. It also involves loosening the tight grip on a paranormal worldview to be able to see the wider expanse of options. A shift in ARIG philosophy can bring about a much-needed improved approach to paranormal investigation that will better serve the needs of individuals and provide useful knowledge to society.

Appendix:
Ghost Hunting Guidebooks

The scientific consensus is that ghosts are not spirits, remnants of the dead, recordings of energy, or supernatural entities. Our existing knowledge about nature does not point to a conclusion that ghosts are a single definable thing, paranormal or normal, that you can find, observe, measure, or study. Yet, from a quick search, I find about 200 guides to "ghost hunting" in print or e-book form that lay out ways to obtain evidence of or contact ghosts. A considerable number of these books have been written by ARIG participants.

It is not independently verifiable that any ghost handbook (or Bigfoot guide book, for that matter) has led anyone to genuinely catch and identify an entity. The supposed value in these books to readers is that they lead you to interpret whatever you find as evidence of an entity. All ghost, monster, and alien hunting instruction books have ultimately failed. But there is good reason to take a closer look at the array of ghost hunting guidebooks. As this volume attests, ghost hunting, as one facet of the paranormal sphere, is an interesting cultural phenomenon, a mainstream hobby, and tourism draw. Paranormal investigation reveals important aspects of society's attitudes towards finding out about the world, deciding what is meaningful and true, using science to examine questions, promoting cooperation and trust in a community, and taking part in a larger effort beyond one's own small role in life. Among other things, by examining the genre of ghost hunting guidebooks, we can get some indication regarding the state of science education (or science literacy) and observe a public discourse about belief and reality.

In examining a selection of several of these books (one person cannot possibly read them all) that span a significant spectrum in time, we can observe the evolution of ghost hunting technique. Many appear to be self-published since several ghost investigation group leaders feel the need to have their own personal volume to use.

A 1973 Practical Guide

Andrew Green's 1973 book *Ghost Hunting: A Practical Guide* is the "first-ever do-it-yourself guide for the psychic researcher." Green was given the snappy moniker of "the Spectre Inspector" in the U.K. and was a well-educated pursuer of ghosts for sixty years. He felt that there was such an interest in the subject of ghosts that there was a need for a small, non-technical guide for the amateur. Green eschews fanaticism and suggests that those interested in the ghost phenomenon study parapsychology, thus reflecting the thinking at that time that academic parapsychology would unlock the mystery of life after death. Therefore, a good portion of the book describes parapsychological concepts, such as telepathy, which he states can be an important consideration as to the cause of a phenomena. He describes Zener cards experiments, which would gain notice after appearing in *Ghostbusters* (1984). This portion of the book will be rather strange to those weaned on 21st century ghost TV shows.

As with many of today's paranormal investigators, Green considered serious ghost hunting important and "groundbreaking" work and the researchers were mavericks. But Green's book contrasts with modern ghost guides, in several ways:

- Green defined ghosts in terms of psi, a parapsychological construct that modern ghost hunters would probably not be familiar with.
- EVPs were called "Raudive voices" and are not emphasized as evidence. Green thought there were too many potential pitfalls to use them this way.
- The technology was primitive compared with what we have today. Equipment included very basic detective-type materials: level, compass, strain-gage, sand or sugar, powder for fingerprints, thread, maybe a camera. But the idea of measuring environmental variables was already being pursued by the Society of Psychical Research at that time.
- Green mentions exorcism but it was clearly not as common as today and people were less bold in advancing the idea.
- Green's advice is that the investigator must be thorough and careful in research and provide a sophisticated investigation. His recommended investigation included studying the geology, geography, and past owners of the location, and producing a report that could be professionally published. This is not on the scale of what is proposed by weekend/overnight trips of today's ghost hunters.

- Green advised the investigator to never (as in *not ever*) get involved in publicity for the case as he recognized that some people are in it just for the attention. Restricting publicity is not what today's investigators typically would agree to since one of their goals is to make a name and reputation for themselves to invite more potential investigations and greater media attention.

One curious "test" that Green proposes is a way to judge the *client* in terms of credentials. This example of using technology is certainly an indication of how much times have changed: "The production of a caseful of apparatus at the commencement of an investigation in itself constitutes a test, for the witness of a genuine phenomena will be, or should be, impressed with the serious nature of ghost hunting, while the fraudulent will be worried by the prospect of being exposed."

Green states "I believe" in his description of how his proposed process should work but few citations of evidential support are provided for these suppositions. For example: Heat extracted from the environment will energize a haunting. Such ideas about ghost manifestations are very old but have yet to be supported or well-argued. Green subscribes to ghosts as real but he is not careless or overtly credulous. He obviously holds expertise in the history of the field, provides many reasonable pieces of sound advice for investigations, and writes at a sophisticated level. This book has been reissued in 2016 by Alan Murdie of The Ghost Club in the U.K. Murdie had recommended this book to me as one of his favorite guides.

Well-Meaning Nonsense, 2003

How to Be a Ghost Hunter by Richard Southall (2003) appears to have been written in 2001 which was the start of the massive proliferation of ghost hunting groups in the U.S. This may have been a "unique handbook" for the time, but it was not for long. The book is of the "Confessions of a Ghost Hunter" type: ghosts are defined, historical aspects are mentioned, prior cases related, procedures and equipment are suggested, collection of data and evidence are described, and advice on forming your own team is offered. Southall states he has a degree in journalism and psychology; the book also has a genuine publisher (although of New Age books), which brings the quality and readability of this guide above most others. However, it follows the typical outline of information and includes many unsupported claims, assumptions and statements of "fact."

Southall, as with most authors of this type of guidebook, assumes that ghosts exists, that paranormal activity is ghost activity, and that these certain descriptions are characteristics of ghosts. How he "knows" this is never explained. No sources are supplied but various unsourced, un-detailed anecdotes substitute as justification. It is assumed the reader will accept these as evidence. Undefined, sciencey-sounding terms are used throughout: "highest amount of paranormal energy," "life force," "psychic energy." He accepts that orbs are an indication of psychic energy. Southall slides into the realm of the supernatural by claiming that if you investigate enough, you will encounter a "demonic entity." Beware, he says, that the Ouija board can invite it in so that device is dangerous to use.

Southall wrote this ghost hunting guide to promote the topic as he was running a ghost tour at the time. He states his role shifted from investigation to teaching. This book fails to bolster the author's scientific credibility when he refers to fictional movies, such as *The Sixth Sense*, to suggest the real world is actually like this. Southall states that the scientific method is the means to get "tangible, measurable evidence" as opposed to psychic impressions and divination, though the two methods can validate each other. He is not a scientist and it shows.

As with Green, the technology used in this time period is quite a contrast to modern investigations. Southall states that "A photograph of a ghost cannot be denied." Considering that faked ghost photos existed since the dawn of photography, this wasn't even rational advice at the time, let alone in the age of today's add-a-ghost phone apps.

He states a good investigator should be unbiased but the language from start to finish is completely biased in the belief that an area is likely haunted. Short shrift is given to examination of mundane causes. But he advises to talk up your own credibility: "Clients love credentials and memberships." The bibliography contains no journals or otherwise scientific sources, but only references to other ghost hunters' books and mass marketed paranormal publications. Southall's writing projects the attitude of a good person who is concerned with people who are having a paranormal problem and want answers that he believes he can provide. He understands that people need reassurance that what they experience is understandable and things will be OK.

Ghost Tech and Science, 2012

The book *Ultimate Ghost Tech* (2012) by Vince Wilson is an example of volume that was used by many ARIGs as part of their education for research and investigation. In one of the forewords (one is spelled "foreword," the other "forword"), Vince is described as the "foremost expert in the technological aspects of paranormal investigation." Note that Wilson is not a scientist or a tech professional. In the other foreword, well-respected parapsychologist Loyd Auerbach blatantly reveals a truism about ghost hunting technology: "Let's face it: ghost hunters love their tech—even if they don't know how to use it or to assess the data from it in light of the reported phenomena." The rest of this book is an example of sounding sciencey but falling short of representing anything like scientific investigation. Wilson's earlier book, *Ghost Science*, was also sloppy—formatted and written with a great many errors. That book begins with a revealing premise: "One of the main purposes of this book is to show that, not only do ghosts exist but also that the laws that govern reality allow them." Wilson's array of three books (3) are essentially self-published and make unsupported claims such as "random energy particles may hold the essence of consciousness…," and "Ghosts will be proven to exist one day and so will psychics.…" He refers to "stuffy scientists" and takes a disparaging tone towards skeptics while using the phrase "just another theory" where "theory" is used to mean "a guess" instead of the scientific meaning of an evidence-supported overarching model of explanation. He apologizes for using mathematical equations to explain a concept in physics, which is odd considering the book is about science and technology. Wilson admits *Ghostbusters* (the movie) changed paranormal research with its lingo and gadgets, "Paranormal research just became really cool overnight." He suggests using science as way to pump up your credibility—not real science, but faking it—saying you should answer questions from people with sciencey words to sound "professional and cool" and a little "nerdy" as people are too embarrassed to ask what you mean.

All ghost hunting guidebooks relate ubiquitous assumptions about ghosts predicated on the belief that they exist (thus scuttling any unbiased investigation of what might really be happening to people). The paradigm of today's ghost investigation is reflected: changes in the environment can be related to ghost behavior and hauntings; technology can provide objective evidence in contrast to just human experience. Wilson suggests that a cold spot could be created (through an explanation of energy transfer) from an entity moving through dimensions. This type of rhetoric (apparent in nearly all ghost hunting guides) is creative but intellectually flimsy. He promises great things for ghost hunters (p. 160): "You can be an amateur parapsychologist and usher in a new era of paranormal research. Wow! That's pretty deep for me!"

Wilson, like many of these guide writers, seems well-meaning, but also willing to learn new things, expand his thinking, and is fairly literate in science ideas—just enough to sound knowledgeable to people who aren't scientists, which is most of the population. Wilson understands that TV ghost hunters are playing a role and that many paranormal investigators are "fooled by an intense need to believe." Hoaxes are rampant. So there is a kernel of truth in much of what he writes. However, that is trumped by his own faith that equipment can detect anomalous energy of some sort. The processes he suggests leave out critical considerations about confounding factors and alternative explanations. Wilson has lectured as a ghost tech expert in the past. He suggests that giving workshops to teach people about this topic is a good way to fundraise for your group.

Sciencey Real Ghost Hunting, 2003

Joshua P. Warren is a fairly well-known personality in on the paranormal scene. His book, *How to Hunt Ghosts* (2003), was produced by an affiliate of Simon and Schuster publishing so the basic elements of a book—grammar, punctuation, spelling, and formatting—are superior to small or self-published efforts. However, the same unsupported model, built on speculative paranormal assumptions, is applied. The first words, "Ghosts are real," show us this is not about investigation but about finding proof to support a preexisting conclusion. These opening words oddly contrast with the last words of the book, "Never pretend to know all the answers. All the answers are not known." In between, there is a mish-mash of sciencey sounding claims and scientific misrepresentation. Warren's resumé does not include science. He writes fiction and worked in film-making. Like many who appear on TV shows, Warren uses these appearances to bolster his credibility. *How to Hunt Ghosts* is replete with scientist namedropping including Descartes, Newton, Einstein, and Sagan.

Warren holds that our current technology is not good enough to easily discover spirits but is oblivious to the reality that physicists can detect subatomic particles and the tiniest perturbations of energy. Yet, our technology is not adequate to identify ghosts?

Warren berates scientists for not paying attention to ghosts because they need to limit their work to activity of a certain category. "Most scientists are busy enough researching the activity they already know about." From the early days of the scientific endeavor, knowledge became specialized by necessity. To say science is flawed because of this is like saying medicine is bad because too many doctors specialize in distinct areas of health or surgery. Specialization is advantageous for advancing deep knowledge.

He states, "Virtually any location can prove to be haunted," and says you should experiment to decide if the Ouija board, automatic writing, pendulums, etc., work for you. However, if *everything* works, then we might logically conclude *nothing* works.

Warren researches natural anomalies like "ghost lights" and runs a "Bermuda Triangle Research" site in Puerto Rico. Therefore, you might possibly put him above "amateur" status since he appears to make a living off this work; it is not professional science but creativity with fringe ideas. His incredible claims about energy fields and "warps" in nature that could explain anomalies do not have equally incredible documentation provided. Like many ghost researchers, Warren knows some science basics but he applies them wildly incorrectly.

Warren is a fiction novelist. He refers to many fictional movies to the point where a reader might question the standards he uses to discern scientific facts from fictional license. Warren provides foundation-less claims that sound like fiction and so fails to support any worthwhile conclusions about ghost experiences. It also leaves these ideas wide open targets for derision by scientists working in legitimate research endeavors. A perusal of his website in 2015 shows that Warren has moved into the business of selling "wishing machines" and lucky charms.

Array of E-Books

I accessed what was a rather random sample of several of the dozens of e-books available online. Unsurprisingly, these also fit into the same template and had similar characteristics:

- "Just so" facts and stories
- No references
- Lack of proofing or editing including several typographical errors and incorrect punctuation
- Poor layout and design
- Unsophisticated, overly casual writing style
- Superficial content

Ultimate Ghost Hunting Guide by Jeff Terrozas, 2011

Subtitled "Everything you need to know for paranormal research," the content is overly rambling and amateurish. Typos abound, the layout is annoyingly sloppy. The premise is that ghost hunting is "fun," so have fun.

It's not to be taken seriously unless you want to make money. In that case, you should act "professional." This book should not be taken seriously.

Ghost Seekers Field Guide, Volume 1
by Frank Potterstone, 2011

No proofreading or editing was apparently done to this manuscript. The language and grammar is poor; typos are abundant and the layout is simply ugly. There is an overuse of ellipses, and random unattributed quotes. Though the author means well, with these factors, the lack of adherence to punctuation conventions, and the unfocused content, this book is unreadable. Yes, there was a Volume 2 as well.

Ultimate Ghost Hunter Field Guide
by Brandy Burgess, n.d.

Layout is very poor with line breaks in the middle of a sentence and random capitalization of words. Grammar is poor and the writing is amateurish and unfocused. The author lays out "facts" such as a description of "psychic burns" and "awakenings" without any support for such supernatural claims. She says you will know a spirit is demonic because of the sulfur or rotten flesh smell as well as the growling sounds. They also appear in half-human, half-animal form. These sound like verifiable claims; one wonders why we can't prove such incredible new findings if they are so obvious.

Ghost Hunting 101: The Ultimate Resource for Beginner and Experienced Ghost Hunters
by Ghostly World, 2015

Ghostly World is a website "dedicated to all things haunted." The authors say on their site that they are not an investigation team or even "in the paranormal field." Yet, here they are publishing and charging for an instruction book on ghost hunting. How's that for credibility? The layout of *Ghost Hunting 101* is acceptable and the writing style is generally appropriate to a serious handbook. There are some typing errors. The content is shallow and lacks development and explanations. Terms and labels are assigned subjectively. For example, readers are told there are three kinds of ghost hunters: a hobbyist, a serious researcher, and a home investigator. A random graph is included (because graphs look sciencey) without any source data to show that 100 percent are hobbyists, 50 percent are serious researchers and only 10 percent are home investigators. Going into a client's home is serious stuff and the ghost hunter needs to provide comfort and assistance to the residents while studying spirits. The unnamed author(s) suggest the ghost hunter may need to act in the capacity of a "therapist"—a highly unethical suggestion. Meanwhile, the reader is warned that Ouija boards and other occult dealings will bring about dangerous evil spirits. They seem to think Grant Wilson and Jason Hawes invented ghost hunting.

Some of these books are surprisingly candid, as I found with *How to Legally Gain Access to Haunted Locations: A Guide for Paranormal Investigators* (n.d.) by Casper Waylin. Waylin makes no apologies for playing pretend and weaseling your way into clients' homes. He recommends following what you see on TV shows:

> Professionalism starts as "pretending" but evolves into something that's real. If you're just getting started as a ghost hunting group, you'll need to pretend that you're a "professional" and put on a convincing act for the people you talk to in order to gain entry into a particular location. Put together a good costume (some nice clothes) and props (legal documents and contracts) and then tell clients and gatekeepers exactly what you plan to do from beginning to end. In terms of how you greet and speak to new clients, it can help to model other group leaders you've seen on TV or read about in books and for crying out loud, make sure that you have a firm handshake and you look them in the eye during your initial contact!
>
> ...
>
> Acting professional is okay if you're not really a professional. Find a character in a movie or watch some of the later episodes of TAPS *Ghost Hunters* or *Ghost Adventures* and emulate the paranormal investigators that you can relate to best.

So copy the guys on TV. Pretend to be like them when you enter other people's houses. This is awful stuff.

Finally, I would like to mention a specialty guide called *The Other Side: A Teen's guide to Ghost Hunting and the Paranormal* (2009) by Gibson, Burns, and Schrader. This might be considered one of the "least worst" books since it was done by a reputable publisher and contains a handful of good advice. There are two overarching and egregious problems with this book: (1) Misinformation directed at teens to take on this topic and "educate the masses" about "what our place is in the universe and what possibilities there are of an afterlife"; and (2) The ignorant and condescending attitude towards science (as difficult and cumbersome) and skepticism (as cynical bullying) (p. 67). The logical fallacies and unsupported claims rampant in this book would make it excellent to use as an example for a critical thinking exercise.

The "Ultimate" Conclusion

Most, perhaps all, of these authors wrote these books because they believed it would be helpful to an audience or to their investigation group to codify what they deemed to be important knowledge and procedures that everyone was expected to follow. With the advent of easy self-publishing, we've seen a proliferation of low-quality, previously unpublishable books like never before. Anyone, even someone who never wrote an article or term paper, can call themselves an "expert," publish a book, and claim to be an author. There is no excuse for publishing a book without having it edited for basic grammar, spelling, and punctuation. There is no justification for the self-serving, misguided misinformation out there that promises the reader that "this book" is the (ultimate) thing you need to set yourself up as a genuine, credible, and successful ghost hunter. My recommendation: Don't bother with any of them. Look up books done by professional science writers or reports by actual parapsychologists to learn the literature of the field before you write a book and say you know what you are talking about.

Scientific Paranormal Investigation by Benjamin Radford (2010) is currently the only rational, skeptical-themed guide out there. If you do any paranormal investigation, this should be on your shelf. Lay parapsychologists should refer to *Parapsychology: A Handbook for the 21st Century* by Cardeña et al. (2015). You can borrow this from a university library or browse it online. While I disagree with some content in this volume, it is an example of a credible way to construct a sophisticated and useful handbook that will be relevant for decades. It will also give the ghost hunter hobbyists an eye-opener on the huge amount of parapsychological research that has been done by far more qualified people of various disciplines. Written at a college reading level, it is not in the same class of books cited above, and makes all amateur guides look extremely unsophisticated in comparison. But if you are going to claim to be doing groundbreaking research that will enhance our future knowledge about spirits and hauntings, you need to up your game. The world does not need any more pro-paranormal ghost hunting guidebooks by self-styled experts.

Chapter Notes

Introduction

1. Work on the "backfire effect" in politics has been done by B. Nyhan. Recommended reading is J. Cook and S. Lewandowski (2011), *Debunking Handbook*, www.skepticalscience.com/docs/DebunkingHandbook.pdf.

2. A snipe hunt is a practical joke played on the young where they are given an impossible task. There are no snipes so those hunting will be looking for them forever and make a fool of themselves.

Chapter 1

1. This is in opposition to Max Weber's idea of "disenchantment" of society via the emphasis of scientific understanding over belief. Re-enchantment tendencies will insert magical meaning into even ordinary life actions.

2. For the TAPS website, visit tapsfamily.com/tf-members/taps-family-members-list.

3. The T.A.P.S. Mentality. Retrieved May 29, 2010, from www.the-atlantic-paranormal-society.com/abouttaps/mentality.html.

4. Blogger Jack Brewer has followed the recent history of MUFON. ufotrail.blogspot.com/2013/07/mufon-science-and-deception-part-one.html.

5. Antonio Paris was a former MUFON member who started his own organization. aerial-phenomenon.org.

Chapter 2

1. www.gallup.com/poll/4483/americans-belief-psychic-paranormal-phenomena-over-last-decade.aspx.

2. www.gallup.com/poll/16915/three-four-americans-believe-paranormal.aspx.

3. Later in this volume, there is an important distinction made between supernatural, which includes religious beliefs, and the paranormal, making this statistic a bit confusing. But it does show the popularity of belief in things beyond mainstream science and knowledge.

4. "Seen any Red Pandas Lately?" www.nsta.org/publications/news/story.aspx?id=53418.

5. www.nytimes.com/2016/09/30/us/creepy-clown-hoaxes-arrests.html.

6. AP created a Twitter account called AP Oddities. *Huffington Post* has a subcategory site called "Weird" as well as a social media account for this. While the topics are broader than ghosts and UFO reports, the popularity of such news outside the mainstream current events news is notable.

7. Check out one of my favorite collections of pop culture kitsch at ilovetheyeti.blogspot.com.

8. The original source for this most wonderful word is undetermined.

9. An example of this is *Ghost Adventures* visiting Rose Hall Plantation in Jamaica regardless of the documentation that the story of the witch of Rose Hall was fictitious.

10. These numbers are from my own list kept up to date at sharonahill.com/paranormaltv.

11. I cite personal experience. I asked vendors of such equipment at a paranormal conference. Many advertisements include the standard "as seen on TV."

12. Paranormal conferences resemble fan conventions more than academic conferences. The television personalities appear for Q&A panels and autograph sessions rather than deliver a lecture or presentation.

13. "Ghost hunters utilize latest in technology / Paranormal research has become a popular pursuit." www.sfgate.com/business/article/Ghost-hunters-utilize-latest-in-technology-2562702.php.

14. "Ghostbusters inspired me to work as a parapsychologist, hunting ghouls." www.theguardian.com/careers/careers-blog/ghostbusters-parapsychologist-spooks.

15. November 17, 2013, "An Interview with Alex Jones, America's Leading (and Proudest) Conspiracy Theorist," *New York*. nymag.com/daily/intelligencer/2013/11/alex-jones-americas-top-conspiracy-theorist.html.
16. See the archives of the Milwaukee Journal Sentinel for the most infamous case of this example of ostension to date—the Slenderman stabbings of 2014. archive.jsonline.com/news/crime/waukesha-slender-man-stabbing-262029261.html.

Chapter 3

1. Description from the Rhine Research Center website www.rhine.org.
2. www.parapsych.org.
3. www.parapsychology.org.
4. See the 2016 *Fortean Times* 338 (April): 16–18.
5. ghostresearch.org/. Accessed from 2010 to 2016.
6. www.ghostweb.com/portal.html.
7. sharonahill.com/paranormaltv.

Chapter 4

1. The origin of the word "cryptid" is, ironically, cryptic Loxton & Prothero 2013: 16–17). Regal (2011:22) attributes it to John Wall in 1893.
2. http://orgoneresearch.com/2009/10/19/florida-"giant-penguin"-hoax-revealed.
3. The first sighting is commonly attributed to Samuel de Champlain but his account is clearly not a monster but a fish.
4. www.skepticblog.org/2013/09/12/breaking-down-a-criticism-of-abominable-science.
5. www.animalplanet.com/tv-shows/finding-bigfoot/bios/matt-moneymaker.
6. North American Wood Ape Conservancy, woodape.org.
7. bionews-tx.com/news/2013/09/01/texas-state-university-researcher-helps-unravel-mystery-of-texas-blue-dog-claimed-to-be-chupacabra.
8. As stated on the website www.olympicproject.com.
9. Chapter 6 of *Abominable Science* details the motivation of some individuals to find Mokele-Mbembe, a living dinosaur, in the African Congo area as proof against evolution.
10. The Sylvanic website periodically disappeared from the Internet and is now officially gone. You can see a report on the photos from Discovery Channel's *Animal Planet* here: blogs.discovery.com/animal_oddities/2011/09/do-new-photos-show-a-real-bigfoot.html.

11. See Cliff's insider's report on this episode on his website: cliffbarackman.com/finding-bigfoot/finding-bigfoot-episode-guide/finding-bigfoot-season-two/finding-bigfoot-season-two-canadian-bigfoot-eh.
12. For several years, I chronicled Bigfoot hoaxes that made the mainstream media news feeds. You can view these stories at doubtfulnews.com/?s=bigfoot+hoax.

Chapter 5

1. www.cufon.org/cufon/robert.htm.
2. www.nuforc.org/General.html.
3. www.cufos.org/org.html.
4. For a review of MUFON's problems over the last few years, I found Jack Brewer's blog "UFO Trail" to be insightful. This post, from July 2013, provides a road map of things that went wrong and why: ufotrail.blogspot.com/2013/07/mufon-science-and-deception-part-one.html.
5. A term coined by social psychologist Leon Festinger in 1957.
6. Contact in the Desert 2015 claimed 2000 attendees. www.kmir.com/story/29202460/contact-in-the-desert-ufo-conference-held-in-joshua-tree.
7. www.openminds.tv/about.

Chapter 6

1. Ohio Paranormal Investigation Network (OPIN).
2. I attended the 2011 Pennsylvania UFO and Bigfoot Conference at Westmoreland Community College.
3. A subset of cryptozoologists entertains the possibility that cryptids are not fully material beings but may be extra-dimensional or spectral entities.
4. Seven men were arrested and pled guilty to trespassing and arson when a fire destroyed the LeBeau Plantation mansion in Louisiana in 2013. They were looking for ghosts. www.nola.com/crime/index.ssf/2014/09/seven_men_who_apparently_were.html.
5. Regal 2011, p. 181.
6. For examples of these situations, research the cases where counselors for purported UFO abductees and satanic ritual abuse victims led their clients to report "recovered memories" as genuine that were later found to be demonstrably false.

Chapter 7

1. Charles Percy Snow (1959), *The Two Cultures* (Cambridge University Press).

2. This speech is readily available online and contained within Feynman's 1985 book *Surely, You're Joking Mr. Feynman*.
3. See the 2007 article "Cargo cult lives on in South Pacific," news.bbc.co.uk/2/hi/asia-pacific/6370991.stm.

Chapter 8

1. en.wikiquote.org/wiki/Samuel_Johnson From Boswell's *Life of Johnson*: "Talking of ghosts, he said ... '[It] is undecided whether or not there has ever been an instance of the spirit of any person appearing after death. All argument is against it; but all belief is for it.'"
2. www.instituteofmetaphysics.com.
3. www.rhine.org.
4. nevadaiops.com.
5. www.flamelcollege.org.
6. www.internationalmetaphysicaluniversity.org.

Chapter 9

1. The Prodigy Paranormal Group.
2. Port City Paranormal.
3. West Coast Paranormal.
4. Wisconsin Paranormal Investigation Team (WIX).
5. Arkansas Paranormal Anomalous Studies Team (ARPAST).
6. The Pennsylvania Paranormal Association (PPA).
7. Central Oklahoma Paranormal Studies.
8. See the November 9, 2011, article "Do Einstein's Laws Prove Ghosts Exist?" at www.livescience.com/16951-einstein-physics-ghosts-proof.html.
9. Researchers of Paranormal Experiences (ROPE).
10. Oregon's Unknown Creepy Haunting and Paranormal Investigations (OUCHPI).
11. East Coast Research and Investigation of the Paranormal.
12. The Ghost Chicks.
13. League of Paranormal Investigators. This "Ghost Lab" device is not the same as that from the *Ghost Lab* television show.
14. Northeast Arkansas Paranormal Society.
15. Druid City Paranormal and Wiregrass Ghost Hunters.
16. East Coast Haunting Org—ECHO Paranormal.
17. National Ghost Hunters Society.
18. Calhoun County Paranormal Investigators.
19. Blue Ridge Paranormal Investigations.
20. Miller's Paranormal Research.
21. Paranormal Investigators of Central Arizona (PICA).
22. International Paranormal Reporting Group.
23. East Coast Haunting Org (ECHO Paranormal).
24. Blue Ridge Paranormal Investigators.
25. Florida Paranormal Investigations.
26. Sonoran Paranormal Investigation, Inc. (SPI).
27. For example, the BRFO database of Bigfoot sightings was used to create a map of sighting distribution in the U.S. and Canada over 92 years. www.joshuastevens.net/visualization/squatch-watch-92-years-of-bigfoot-sightings-in-us-and-canada.
28. The ARIGs I contacted were not keen on sharing these views publicly. They seemed intent on protecting their existing reputation. I am grateful for their candor with me.

Chapter 10

1. Paranormal Investigators of Central Arizona (PICA).
2. skepdic.com/pareidol.html.
3. The "lights out" concept seems counterintuitive to investigations in general. Why would ghosts only appear in the dark? Various reasons for this have been stated by researchers. For one, the entities could be repelled by the light. Other researchers claim that they are removing sources of electromagnetic radiation that may interfere with the collection of data. Those and all other reasons are questionable, but well-established is that "lights out" is conducive to a more edgy environment that heightens perceptions (and misperceptions).
4. Standardized image shared between websites.
5. See the 2015 "The Ouchita Project Report," March, NAWAC. media.texasbigfoot.com/OP_paper_media/OuachitaProjectMonograph_Version1.1_03112015.pdf.
6. See "The Haunted Bed" study at www.assap.ac.uk/newsite/htmlfiles/MADS haunted bed.html.
7. www.scientificamerican.com/article/ghost-lusters-if-you-want/.
8. Wiseman, et al., 2002, "An investigation into the alleged haunting of Hampton Court Palace: Psychological variables and magnetic fields," *Journal of Parapsychology* 66(4), 387–408.
9. The movement of the planchette has been demonstrated to be due to the ideomotor effect whereby people unwittingly move the device via small muscular movements.

10. Searching Out and Uncovering Lost Spirits (SOULS).
11. 6 Cents Investigations.
12. "Exorcisms On the Rise," www.catholiceducation.org/en/culture/catholic-contributions/exorcisms-on-the-rise.html.
13. See the review of this program at variety.com/2015/tv/reviews/exorcism-live-review-destination-america-the-exorcist-house-1201629070.
14. See information on "Xenu" (en.wikipedia.org/wiki/Xenu), dictator of the Galactic Federation.
15. See the list of UFO religions on Wikipedia, en.wikipedia.org/wiki/List_of_UFO_religions.
16. Spirit Light Network.
17. After Midnight Paranormal Investigation Team.

Chapter 11

1. In 2008, leading researchers in parapsychology Robert G. Jahn and Brenda J. Dunne of the Princeton Engineering Anomalies Research (PEAR) wrote a paper called "Change the Rules" advocating just this.
2. This idea that one should not attempt to explain something until you are sure there is something to be explained has been referred to as a "skepticalism" and credited to psychologist Ray Hyman and sometimes called "Hyman's Categorical Imperative" or "Hyman's Maxim." www.skeptic.com/insight/history-and-hymans-maxim-part-one
3. Triangle Paranormal Investigations (TPI).
4. REAL Paranormal Investigators.
5. City Lights Paranormal Society.
6. The Prodigy Paranormal Group.
7. Southeast Idaho Paranormal Organization.

Chapter 12

1. See the 2014 results here: www.nsf.gov/statistics/seind14/index.cfm/chapter-7/c7s2.htm.
2. NSF survey regarding teaching of evolution: https://www.nsf.gov/statistics/seind14/index.cfm/chapter-7/c7s4.htm#s7.
3. johnemackinstitute.org.
4. www.nytimes.com/1995/05/04/us/harvard-investigates-a-professor-who-wrote-of-space-aliens.html.
5. See www.csicop.org/specialarticles/show/supernatural_creep_the_slippery_slope_to_unfalsifiability.
6. This idea was promoted in response to the Ketchum DNA results and Bigfoot researcher Ron Morehead. ronmorehead.com/the-nephilim-theory.
7. Paranormal Analysis and Research Association (PARA).

Conclusion

1. This section on Bobby and Jason was originally published as "Giving up the Ghosts: Formerly Known as 'Ghost Hunters'" at csicop.org/specialarticles/show/givinguptheghostsformerlyknownasghosthunters in April 2014 in my online column called "Sounds Sciencey."
2. The article "Two Men's Quest for Florida's Mysterious Skunk Ape" can be found at www.browardpalmbeach.com/news/two-mens-quest-for-floridas-mysterious-skunk-ape-7966175.
3. It remains confusing to me why someone would think I would undertake such a dangerous venture (alligators, mosquitos, and all) in pursuit of a creature that does not plausibly exist. I leave zoology to the professionals whose job it is to catalog the diversity of life.
4. I highly recommend *Not Your Typical Bigfoot Movie* (2008), www.imdb.com/title/tt1185384/, which expresses the same undertones of friendship, hope, and escapism in another pair of amateur Bigfooters.
5. See the 1990 article "Why We Need to Understand Science," *Skeptical Inquirer* 14.3, Spring. Available at www.csicop.org/si/show/why_we_need_to_understand_science.
6. Benjamin Radford, 2002, "Bigfoot at 50 Evaluating a Half-Century of Bigfoot Evidence," *Skeptical Inquirer* 26.2, March/April. Available at www.csicop.org/si/show/bigfoot_at_50_evaluating_a_half-century_of_bigfoot_evidence.

References

Agin, D. (2006). *Junk Science*. Thomas Dunne Books.

Allison, P. D. (1979). "Experimental Parapsychology as Rejected Science." In R. Wallis (Ed.), *On the Margins of Science: The Social Construction of Rejected Knowledge*, pp. 271–291.

Andrews, D. (2007). *A Challenge to Science Education: How Do Student and Teacher Beliefs in the Paranormal and Pseudoscience Affect Scientific Literacy?* [Thesis] Masters in Education in Curriculum and Instruction, Pennsylvania State University.

Bader, C. D., F. C. Mencken, and J. O. Baker. (2010). *Paranormal America*. New York University Press.

Baker, R. A., and J. Nickell. (1992). *Missing Pieces: How to Investigate Ghosts, Ufos, Psychics, & Other Mysteries*. Prometheus Books.

Bartholomew, R. E. (2012). *The Untold Story of Champ*. State University of New York Press.

Belz, M., and W. Fach. (2015). "Exceptional Experiences (ExE) in Clinical Psychology." In Cardena, et al. (Eds.), *Parapsychology: A Handbook for the 21st Century*, pp. 364–379.

Beveridge, W. I. B. (1957). *The Art of Scientific Investigation*, rev ed. W. W. Norton & Co., Inc.

Blake, J. A. (1979). "The Intellectual Development and Social Context of the Study of Unidentified Flying Objects." In R. Wallis (Ed.), *On the Margins of Science*, pp. 315–337.

Blancke, S., M. Boudry, and M. Pigliucci (2016). "Why Do Irrational Beliefs Mimic Science? The Cultural Evolution of Pseudoscience." *Theoria*. doi:10.1111/theo.12109

Blum, D. (2006). *Ghost Hunters*. Penguin Books.

Booker, M. K. (2009). *Red White and Spooked: The Supernatural in American Culture*. Praeger.

Brown, A. (2008). *Ghost Hunters of New England*. University Press of New England.

Buhs, J.B. (2009). *Bigfoot: The Life and Times of a Legend*. University of Chicago Press.

Bunge, M. (1984). "What Is Pseudoscience?" *Skeptical Inquirer* 9 (Fall), 36–46.

Cardena, E. (2015). "On Negative Capability and Parapsychology: Personal Reflections." In Cardena, et al. (Eds.), *Parapsychology: A Handbook for the 21st Century*, pp. 399–404.

Cardena, E., J. Palmer, and D. Marcusson-Clavertz, eds. (2015). *Parapsychology: A Handbook for the 21st Century*. McFarland.

Childs, C., and C. D. Murray. (2010). "'We All Had An Experience In There Together': A Discursive Psychological Analysis of Collaboration." *Qualitative Research in Psychology* 7, 21–33.

Clarke, D. (2013). "Extraordinary Experiences with UFOs." In Jenzen and Munt (Eds.), *The Ashgate Research Companion to Paranormal Cultures*, pp. 79–94.

Clarke, R. (2012). *A Natural History of Ghosts*. Penguin Books.

Collins, H. M. (1987). "Certainty and the Public Understanding of Science: Science on Television." *Social Studies of Science* 17(4), 689–713.

Collins, H. M. (2006). "Lone Voices Special: How We Know What We Know." *New Scientist* 2581 (9 December 2006).

Collins, H. M., and T. J. Pinch. (1982). *Frames of Meaning: The Social Construction of *Extraordinary Science*. Routledge & K. Paul.

Daniels, G. H. (1971). *Science in American Society*. Knopf.

Davies, O. (2007). *The Haunted: A Social History of Ghosts*. Palgrave Macmillan.

Degele, N. (2005). "On the Margins of Everything: Doing, Performing, and Staging Science in Homeopathy." *Science Technology Human Values* 30(1), 111–136.

Denzler, B. (2003). *The Lure of the Edge*. University of California Press.

Derksen, A. A. (1993). "The Seven Sins of Pseudo-Science." *Journal for General Philosophy of Science* 24(1), 17–42.

Dewan, W. J. (2013). "Sceptic Culture: Traditions of Disbelief in New Mexico." In Jenzen and Munt (Eds.), *The Ashgate Research Companion to Paranormal Cultures*, pp. 107–120.

Dewitt, R. (2004). *Worldviews: An Introduction to the History and Philosophy of Science*. Blackwell.

Diamandis, E. P. (2013). "Nobelitis: a common disease among Nobel laureates?" *Clinical Chemistry and Labaratory Medicine* 51(8), 1573–1574.

Dobry, D. (2013). "Interpreting Death and the Afterlife in USU.S. Paranormal Reality Television Programmes and Online Fan Groups." In Jenzen and Munt (Eds.), *The Ashgate Research Companion to Paranormal Cultures*, pp. 171–188.

Dodds, R., E. Tseelon, and E. L. C. Weitkemp. (2005). "Making Sense of Scientific Claims in Advertising: A Study of Scientifically Aware Consumers." *Public Understanding of Science* 17, 211–230.

Dolby, R. G. A. (1975). "What Can We Usefully Learn from the Velikovsky Affair?" *Social Studies of Science* 5(2), 165–175.

Dolby, R. G. A. (1979). "Reflections on Deviant Science." In R. Wallis (Ed.), *On the Margins of Science: The Social Construction of Rejected Knowledge*, pp. 9–47.

Duffy, R. (2012). *Survey of Paranormal Research and Ghost Hunting Groups*. Independent Investigations Group, Colorado.

Edwards, E. D. (2001). "A House That Tries to Be Haunted: Ghostly Narratives in Popular Film and Television." In J. Houran and R. Lange (Eds.), *Hauntings and Poltergeists: Multidisciplinary Perspectives*, pp. 82–120.

Eghigian, G. (2014). "A Transatlantic Buzz: Flying Saucers, Extraterrestrials and Americans in Postwar Germany." *Journal of Transatlantic Studies* 12(3), 282–303.

Eghigian, G. (2015). "Making UFOs Make Sense: Ufology, Science and the History of Their Mutual Mistrust." *Public Understanding of Science* (online), December 6. http://journals.sagepub.com/doi/abs/10.1177/0963662515617706.

Finucane, R. C. (1996). *Ghosts: Appearances of the Dead and Cultural Transformation*. Prometheus Books.

Friedlander, M. W. (1995). *At the Fringes of Science*. Westview Press.

Gardner, M. (1957). *Fads and Fallacies in the Name of Science*, 2nd ed. Dover Publications.

Gauchat, G. (2010). "The Cultural Authority of Science: Public Trust and Acceptance of Organized Science." *Public Understanding of Science*. Published online before print May 27, 2010. http://journals.sagepub.com/doi/abs/10.1177/0963662510365246.

Gibson, M., P. Burns, and D. Schrader. (2009). *The Other Side: A Teen's Guide to Ghost Hunting and the Paranormal*. Houghton Mifflin Harcourt.

Gigerenzer, G. (2009). "Surrogates for Theory." *Observer* 22(2).

Gieryn, T. F. (1983). "Boundary-Work and the Demarcation of Science from Non-Science: Strains and Interests in Professional Ideologies of Scientists." *American Sociological Review* 48(6), 781–795.

Goode, E. (2000). *Paranormal Beliefs: A Sociological Introduction*. Waveland Press.

Gordin, Michael D. (2012). *The Pseudoscience Wars: Immanuel Velikovsky and the Birth of the Modern Fringe*. University of Chicago Press.

Gordon, S. (2010). *Silent Invasion: The Pennsylvania UFO-Bigfoot Casebook*. Self-published.

Gordon, D. G. (2015). *The Sasquatch Seekers Field Manual*. Mountaineer Books.

Gregory, J., and S. Miller. (2000). *Science in Public: Communication, Culture, and Credibility*. Perseus.

Grinspoon, L., and A. D. Persky. (1972). "Psychiatry and UFO Reports." In Sagan and Page (Eds.), *Ufo's: A Scientific Debate*, pp. 233–246.

Haack, S. (1997). "Science, Scientism, and Anti-Science in the Age of Preposterism" [Electronic Version]. *Skeptical Inquirer* 21. http://www.csisop.org/si/show/science_scientism_and_anti-science_in_the_age_of_preposterism/.

Haack, S. (2003). *Defending Science—within Reason: Between Scientism and Cynicism.* Prometheus Books.

Haard, J., M. D. Slater, and M. Long. (2004). "Scientiese and Ambiguous Citations in the Selling of Unproven Medical Treatments." *Health Commununication* 16(4), 411–426.

Hall, R. L. (1972). "Sociological Perspectives on UFO Reports." In Sagan and Page (Eds.), *Ufo's: A Scientific Debate*, pp 213–223.

Harvey, J. (2013). "The Ghost in the Machine: Spirits and Technology." In Jenzen and Munt (Eds.), *The Ashgate Research Companion to Paranormal Cultures*, pp. 51–64.

Haythornthwaite, C., and L. Kendall. (2010). "Internet and Community." *American Behavioral Scientist* 53(8), 1083–1094.

Hess, D. J. (1993). *Science in the New Age: The Paranormal, Its Defenders and Debunkers, and American Culture.* University of Wisconsin Press.

Hill, A. (2010). *Paranormal Media.* Routledge.

Hill, A. (2013). "Paranormal Cultural Practices." In Jenzen and Munt (Eds.), *The Ashgate Research Companion to Paranormal Cultures*, pp. 65–78.

Hill, S. (2010). *Being Scientifical: Popularity, Purpose and Promotion of Amateur Research and Investigation Groups in the U.S.* [Thesis] Master of Education, University at Buffalo, State University of New York.

Hill, S. (2013). "The Ketchum Project: What to Believe about Bigfoot DNA 'Science.'" *Skeptical Briefs* 23(1). http://www.csicop.org/sb/show/the_ketchum_project_what_to_believe_about_bigfoot_dna_science.

Hines, T. (2003). *Pseudoscience and the Paranormal*, 2nd ed. Prometheus Books.

Houran, J., and R. Lange (Eds.). (2001). *Hauntings and Poltergeists: Multidisciplinary Perspectives.* McFarland.

Hufford, D. J. (2001). "An Experimental Approach to Hauntings." In J. Houran and R. Lange (Eds.), *Hauntings and Poltergeists: Multidisciplinary Perspectives*, pp. 18–40.

Hynek, J. A. (1972). "Twenty-One Years of UFO Research." In Sagan & Page, *Ufo's: A Scientific Debate*, pp 37–51.

Irwin, H. J. (1989). *Introduction to Parapsychology.* McFarland (5th ed., 2007).

Jenzen, O., and S. R. Munt (Eds.). (2013). *The Ashgate Research Companion to Paranormal Cultures.* Ashgate Publishing.

Keel, J. A. (1975). "The Flying Saucer Subculture." *J. of Popular Culture* 8(4), 871–896.

Kelly, E. W., and J. B. Tucker. (2015). "Research Methods with Spontaneous Case Studies." In Cardena, et al. (Eds.), *Parapsychology: A Handbook for the 21st Century*, pp. 63–76.

Ketchum, M. S., et al. (2013). "Novel North American Hominins, Next Generation Sequencing of Three Whole Genomes and Associated Studies." *DeNovo*—Special Edition, 13 February. http://www.denovojournal.com/denovo_002.htm.

Kleif, T., and W. Faulkner (2003). "'I'm No Athlete [but] I Can Make This Thing Dance!'—Men's Pleasures in Technology." *Science Technology Human Values* 28(2), 296–325.

Koven, M. J. (2007). "Most Haunted and the Convergence of Traditional Belief and Popular Television." *Folklore* 118(2), 183–203.

Krulos, T. (2015). *Monster Hunters.* Chicago Review Press.

Lamont, P. (2007). "Paranormal Belief and The Avowal of Prior Skepticism." *Theory & Psychology* 17(5), 681–696.

Lankford, J. (1981). "Amateurs and Astrophysics: A Neglected Aspect in the Development of a Scientific Specialty." *Social Studies of Science* 11(3), 275–303.

Leary, M. R. and T. Butler. "Electronic Voice Phenomena." In Cardena, et al. (Eds.), *Parapsychology: A Handbook for the 21st Century*, pp. 341–349.

Ley, W. (1968). *Dawn of Zoology.* Prentice-Hall.

License, T. (2016). "The Ghosts of the Ostrich Inn." *Fortean Times* 342: 42–46.

Loftus, E. F. (1996). *Eyewitness Testimony.* Harvard University Press.

Long, G. (2004). *The Making of Bigfoot.* Prometheus Books.

Loxton, D. and D. Prothero (2013). *Abominable Science.* Columbia University Press.

Lyons, S. L. (2009). *Species, Serpents, Spir-*

its and Skulls. State University of New York Press.

Maddox, B. (2009). "The Quick and The Dead." *Discover* 30(7), 72–74.

Maher, M. (2015). "Ghosts and Poltergeists: An Eternal Enigma." In Cardena, et al. (Eds.), *Parapsychology: A Handbook for the 21st Century*, pp. 327–340.

Markovsky, B. N., and S. R. Thye (2001). "Social Influence on Paranormal Beliefs." *Sociological Perspectives* 44(1), 21–44.

Marks, D. F. (1986). "Investigating the Paranormal." *Nature* 320 (13 March), 119–124.

Merton, R.K. (1942). "Science and Technology in a Democratic Order." *Journal of Legal and Political Sociology* 1, 115–126.

Mayer, G. (2003). "A Phenomenology of the Ghosthunting Scene in the USA and Germany." In Jenzen and Munt (Eds.), *The Ashgate Research Companion to Paranormal Cultures*, pp. 363–376.

McCue, P. A. (2002). "Theories of Hauntings: A Critical Review." *Journal of the Sociological of Psychical Research* 661(866), 1–21.

McDonald, J. E. (1972). "Science in Default: Twenty-two Years of Inadequate UFO Investigations." In Sagan and Page (Eds.), *Ufo's: A Scientific Debate*, pp. 52–122.

McRae, M. (2012). *Tribal Science*. Prometheus Books.

Meurger, M. (1989). *Lake Monster Traditions: A Cross-Cultural Analysis*. Fortean Tomes—The Book Service, Ltd.

Michael, M. (1992). "Lay Discourse of Science: Science-in-General, Science-in-Particular, and Self." *Science, Technology and Human Values* 17(3), 313–333.

Miller, B. (2015). "Quantum Theory and Parapsychology." In Cardeña, et al. (Eds.), *Parapsychology: A Handbook for the 21st Century*, pp. 165–180.

Mims, F. M. (1999). "Amateur Science—Strong Tradition, Bright Future." *Science* 284(5411), 55–56.

Molle, A., C.D. Bader (2013). "'Paranormal Science' from America to Italy: A Case of Cultural Homogenization." In Jenzen and Munt (Eds.), *The Ashgate Research Companion to Paranormal Cultures*, pp. 121–137.

Mooney, C., and S. Kirshenbaum. (2009). *Unscientific America: How Scientific Illiteracy Threatens our Future*. Basic Books.

Morrison, P. (1972). "The Nature of Scientific Evidence: A Summary." In Sagan and Page (Eds.), *Ufo's: A Scientific Debate*, pp. 276–290.

Moseley, J. W. and K. T. Flock (2002). *Shockingly Close to the Truth*. Prometheus Books.

Naish, D. (2016). *Hunting Monsters*. Arcturus Books.

National Science Foundation (NSF). (2009). "Science and Engineering Indicators 2010." National Science Board (NSB 10-01). Available via http://www.nsf.gov/statistics/seind10/start.htm.

Nelkin, D. (1987). *Selling Science*. W.H. Freeman & Co.

Nickell, J. (2012). *CSI Paranormal: Investigating Strange Mysteries*. Inquiry Press.

Northcote, J. (2007). *The Paranormal and the Politics of Truth: A Sociological Account*. Imprint Academic.

O'Connor, J. G., and A. J. Meadows. (1976). "Specialization and Professionalization in British Geology." *Social Studies of Science* 6(1), 77–89.

Partridge, C. (2013). "Haunted Culture: The Persistence of Belief in the Paranormal." In Jenzen and Munt (Eds.), *The Ashgate Research Companion to Paranormal Cultures*, pp. 39–50.

Peirce, C. S. (1931). "Morality and Sham Reasoning." *The Collected Papers, Vol 1: Principles of Philosophy* (Section 6). http://www.textlog.de/peirce_principles.html.

Pew Research Center. (2009). *Eastern, New Age Beliefs Widespread: Many Americans Mix Multiple Faiths*. Pew Forum on Religion and Public Life. http://pewforum.org/Other-Beliefs-and-Practices/Many-Americans-Mix-Multiple- Faiths.aspx.

Pigliucci, M. (2010). *Nonsense on Stilts: How to Tell Science From Bunk*. University of Chicago Press.

Pigliucci, M., and M. Boudry (Eds.). (2013). *Philosophy of Pseudoscience*. University of Chicago Press.

Pitrelli, N., F. Manzoli, and B. Montolli. (2006). "Science in Advertising: Uses and Consumptions in the Italian Press." *Public Understanding of Science* 15(2) 207–220.

Potts, J. (2004). "Ghost Hunting in the Twenty-First Century." In J. Houran (Ed.), *From Shaman to Scientist: Essays on Humanity's Search for Spirits*, pp. 211–232.

References

Radford, B. (2010). *Scientific Paranormal Investigation: How to Solve Unexplained Mysteries*. Rhombus Publishing.

Radford, B., and J. Nickell (2006). *Lake Monster Mysteries*. University Press of Kentucky.

Regal, B. (2011). *Searching for Sasquatch: Crackpots, Eggheads and Cryptozoology*. Palgrave McMillan.

Ridolfo, H., A. Baxter, and J. W. Lucas. (2010). "Social Influences on Paranormal Belief: Popular Versus Scientific Support." *Current Research in Social Psychology* 15(3).

Sagan, C. (1972). "UFO's: The Extraterrestrial and Other Hypotheses." In Sagan and Page (Eds.), *Ufo's: A Scientific Debate*, pp. 265–275.

Sagan, C., and T. Page (Eds.). (1972). *Ufo's: A Scientific Debate*. W.W. Norton & Co.

Sconce, J. (2000). *Haunted Media: Electronic Presence from Telegraphy to Television*. Duke University Press.

Seidman, R. (2009). "Ghost Hunters Continues Ratings Success" [Electronic Version]. *TV by the Numbers*. Retrieved May 29, 2010, from http://tvbythenumbers.com/2009/09/04/ghost-hunters-continues-ratings-success/26220.

Sheaffer, R. (1986). *The UFO Verdict: Examining the Evidence*. Prometheus Books.

Sheaffer, R. (2016). "Missiles and Saucers." *Skeptical Inquirer* 40(2), 22–23.

Shirky, C. (2008). *Here Comes Everybody: The Power of Organizing without Organizations*. Penguin Press.

Sparks, G. G., and W. Miller. (2001). "Investigating the Relationship Between Exposure to Television Programs That Depict Paranormal Phenomena and Beliefs in the Paranormal." *Communication Monographs* 68(1), 98–113.

Stebbins, R. A. (1982). "Serious Leisure: A Conceptual Statement." *Pacific Sociology Review* 25(2), 251–272.

Stebbins, R. A. (1992). *Amateurs, Professionals, and Serious Leisure*. McGill-Queen University Press.

Stebbins, R. A. (2007). *Serious Leisure: A Perspective for Our Time*. Transaction Publishers.

Stoeber, M. and H. Meynell. (Eds.) (1996). *Critical Reflections on the Paranormal*. State University of New York Press.

Sullivan, W. (1972). "Influence of the Press and Other Mass Media." In Sagan and Page (Eds.), *Ufo's: A Scientific Debate*, pp. 258–264.

Sykes, B. C., R. A. Mullis, C. Hagenmuller, T. W. Melton, and M. Sartori, et al. (2014). "Genetic Analysis of Hair Samples Attributed to Yeti, Bigfoot and Other Anomalous Primates." *Proceedings of the Royal Society* 281 (1789). http://rspb.royalsocietypublishing.org/content/281/1789/20140161.

Sykes, B. C. (2016). *Bigfoot, Yeti and the Last Neanderthal*. Disinformation Books.

Thurs, D. P. (2007). *Science Talk: Changing Notions of Science in American Popular Culture*. Rutgers University Press.

Toumey, C. P. (1996). *Conjuring Science: Scientific Symbols and Cultural Meanings in American Life*. Rutgers University Press.

Truzzi, M. (1998). "The Perspective of Anomalistics." http://www.skepticalinvestigations.org/anomalistics/perspective.htm.

Von Lucadou, W. and F. Wald. (2014). "Extraordinary Experiences in its Cultural and Theoretical Context." *International Review of Psychiatry* 26(3), 324–334.

Wallis, R. (Ed.). (1979). "On the Margins of Science: The Social Construction of Rejected Knowledge." *Sociological Review*, Monograph 27. University of Keele.

Walter, B. S. Gardenour. (2013). "Phantasmic Science: Medieval Theology, Victorian Spiritualism, and the Specific Rationality of Twenty-First Century Ghost Hunting." *Jefferson Journal of Science and Culture* 2, 133–148.

Warren, J. P. (2003). *How to Hunt Ghosts: A Practical Guide*. Fireside.

Westrum, R. (1975). "A Note on Monsters." *Journal of Popular Culture* 8(4), 862–870.

Westrum, R. (1977). "Social Intelligence About Anomalies: The Case of Ufos." *Social Studies of Science* 7(3), 271–302.

Westrum, R. (1978). "Science and Social Intelligence About Anomalies: The Case of Meteorites." *Social Studies of Science* 8(4), 461–493.

Wilson, V. (2008). *Ghost Science*. Cosmic Pantheon Press.

Wilson, V. (2012). *Ultimate Ghost Tech*. Cosmic Pantheon Press.

Wiseman, R. (2011). *Paranormality*. Spin Solutions, Ltd.

Ziman, J. M. (2000). *Real Science: What It Is, and What It Means*. Cambridge University Press.

Index

abductees 74–75, 168, 232
Abominable Science 122, 207, 215
acronyms 74, 86, 118
Adamski, George 74
Aerial Phenomenon Research Organizations (APRO) 70
Agassiz, Louis 57
alien abduction 32, 69, 72, 74–75, 79, 98, 120, 128, 168, 194; "grays" 32, 79, 150
alien implants 118, 167, 168
aliens, psychic 173
Almasty 59, 130
Amalgamated Saucer Clubs of America 74
amateur research and investigation groups: categories 23–24; criteria 15–16; demographics 20, 52, 87, 89; mission 25–26, 92, 97, 175, 214, 223; motivation 87, 97, 182, 194, 201, 204, 205, 207; number 19–20; organization 19–21, 63, 72, 111, 118, 120; public appeal 121, 123, 129, 134, 153, 208; purpose 6, 16, 29, 86, 98, 128, 186, 206; social aspects 96–99, 128, 201, 206, 207, 208; values 125–126, 186, 214, 217; *see also* ARIGs
amateurs 2, 3, 4, 11, 13, 17, 18, 21–22, 40, 49, 60, 81, 96, 108, 120, 123–124, 129, 208; and professionals 18, 21, 57, 60, 90, 91, 109, 121, 130–131, 172, 176, 190, 222, 227; vs. scientists and academia 11, 18, 22, 24, 60, 93, 100, 109, 110, 121, 129, 146, 147
Amazing Stories 79
American Association for Advancement of Science (AAAS) 47, 73
American Ghost Society 54
American Institute of Metaphysics 124
American Society of Psychical Research *see* Society of Psychical Research, American
Ancient Aliens 80
anecdotes 9, 35, 64, 66, 120, 135, 142, 144, 157–159, 190, 218, 224
angels 147, 176, 178, 196, 219
anomalistics 193
anomalous cognition 48, 49, 196; *see also* ESP; clairvoyance; psi; psychics/mediums
anomaly hunting 33, 78, 79, 95, 131, 136, 144, 160, 162, 170, 181–183, 214, 220

anthropologists 58, 60, 63, 65, 116, 178
anti-science 25, 34–35, 58, 172
apophenia *see* pareidolia
apparitions 7, 49, 169, 170–171
archaeology (pseudo) 79–80
Area 51 77
argot 118, 208
ARIGs: clients with mental issues 98–99, 229; college affiliation 54, 93; criticism 33, 55, 77, 87, 122–123, 183, 191, 199, 205, 210, 214, 216, 220 (*see also* paranormal unity); educating the public 54–55, 86, 92–96, 154, 186, 224, 229; depicting science 24, 25–26, 39, 49, 81, 113, 118, 132–155, 185–186, 187, 191, 202, 210, 212, 224, 225–226 (*see also* sciencey; scientifical; scientificity); expressing certainty 138, 186, 224; harm/hazards 97, 205; helping the public 16, 52, 75, 86, 92, 111, 135, 203; image 86–87, 108, 208 (*see also* hero); legal liability 99, 229; media and 37, 38, 39, 40, 43, 54–55, 87, 89, 93, 111, 118, 127, 134–136; methods 31, 72, 122, 134, 140–146, 153–155, 157, 172–173, 180, 181, 187, 191, 202; naming 86 (*see also* acronyms); occult and religious practices 142, 172–179, 195, 196, 212; paranormal assumptions 51, 144, 165, 167, 182–184, 219, 225, 226 (*see also* bias; sham inquiry); playing pretend scientist 117, 131, 133, 137, 141, 154, 155, 188, 202, 207, 212, 229; professional/professionalism 13, 25, 47, 54, 55, 89–92, 121, 124, 131, 133, 155, 160, 229; publicity 87, 129, 223; qualifications 55, 85–86, 92, 94, 95, 99, 105, 132, 139, 202, 230; relation to service jobs 52, 53, 86, 90, 179; results, documenting 21, 48, 53, 85, 86, 133, 142, 144–146, 154, 155, 156, 213, 220, 222; scholarship 73, 95, 124, 144–145, 158, 208, 218 222; scientific community and 11, 18, 24–25, 55, 71, 86, 108, 111, 120–123, 128, 129, 133–134, 143, 154–155, 187, 195; skepticism 10, 33, 77, 158, 187, 190, 198–199, 203, 204, 208, 214, 215, 216, 225, 229; television influence 54, 62, 129, 134, 142, 201, 203, 204 (*see also* paranormal television); training 16, 23, 52, 54, 72, 93, 133, 140, 220; use of science

241

132–155 (scientific jargon 118, 136–139, 146, 203; scientific method 12, 132, 140–142, 154, 191, 202, 209, 224); technology (gadgets) 5, 25, 39, 41, 53, 54, 78, 81–84, 86, 89, 97, 98, 133, 135, 136, 138, 140–142, 143, 156, 158, 159–160, 165, 172, 202, 218, 222, 223, 224, 225, 226 (*see also* sciencey; scientifical; scientificity); vs. paranormal investigator 31; vs. parapsychologists 48, 49, 55, 92, 93, 129, 166, 172, 190; vs. scientists and academics 13, 15, 43, 55, 63, 73, 90, 100, 101, 105, 118, 110, 111, 107–108, 118, 120, 121–123, 124, 129, 130–131, 132–134, 138, 187, 190, 208, 209–212, 214, 225, 226; vs. skeptics 33, 85, 89, 122, 159, 181, 183, 184–185, 197–199, 206, 209, 214, 225
ARPAST 146, 154–155, 233
The Art of Scientific Investigation 184
Asia 58, 59, 130
ASSAP 129
astrology 28, 34, 125, 192
atmospheric conditions 171–172; *see also* environmental factors
audio recording 140, 155, 156, 165, 166, 182; *see also* equipment, audio; evidence, audio
Auerbach, Loyd 50, 225

backfire effect 9, 231
Bagans, Zak 20; *see also* Ghost Adventures
balloons (as UFOs) 7 6, 150–151, 164
Baylor Religion Surveys 2, 28
bears 131
Beckjord, Jan-Erik 64
belief: dangers 92, 98, 108, 134, 176, 177; media contribution 41–42, 76, 119; reinforcing 16, 43, 96, 111, 139, 182, 186, 195–196, 201, 205
belief buddies 201–206
believers vs. non-believers 158–159, 199; *see also* skepticism
bestiaries 56
bias 1, 10, 66, 103, 112, 133, 143, 144, 157, 181–184, 186, 187, 198, 210, 215–218, 224, 225; confirmation 182, 184
Bible 66, 80, 126, 196; *see also* religion
Biddle, Kenny 7, 24, 36, 47, 72, 78, 82, 88, 149, 161, 162, 193, 204
Bigfoot 1, 11, 16–17, 19, 21, 22, 23, 28, 29, 31, 32, 35, 40, 44, 56–67, 75, 79, 80, 81, 84, 85, 96, 87, 89, 91, 92, 93, 95–99, 102, 110–111, 120, 123, 125, 127, 129, 130–131, 134, 136, 138, 145, 146, 151, 152, 159, 161, 162, 164, 166–167, 171, 172, 173, 176–178, 183, 186, 191, 192, 193, 195–196, 205, 211, 221; communication 166 (*see also* wood knocking); flesh and blood 29, 63, 66, 95; Native Americans 56, 58, 151, 172, 176, 177, 178; psychic 64, 75, 173, 177, 196; remains (none) 59, 63–64, 91; spirit 29, 65, 125, 151, 172, 175–176, 177, 178, 196; supernatural 64, 173, 177, 178, 195–196; UFOs 29, 32, 64, 75, 89, 196
Bigfoot Field Research Organization (BFRO) 63, 146

Bigfootology 188
biological sample kit 160
Biscardi, Tom 67
black panthers *see* cats
blinded (experiments) 143, 171, 173, 181, 182
blobsquatch 152, 161, 183
blue dogs 64, 168; *see also* chupacabra
Borley Rectory 50
Brewer, Jack 231, 232
Buell, Ryan 55, 135; *see also* Paranormal State
BUFORA 70
burial grounds 51, 86
burns 167, 228
Burns, John 58

camera glitches 148–149, 151, 160, 162, 170, 219
cameras 160–163; FLIR (infrared) 81, 159, 160, 163, 202; night vision 37, 152, 159, 163, 196; remote 59, 63, 81, 161, 162; video 163–164
Canada 20, 233
Cardena, Edzel 47, 48, 129, 190
cargo cult science 117, 233
cats (big, out of place) 64, 151
celebrities (paranormal) 12, 14, 27, 39–40, 53, 89, 91, 124, 173
cemeteries *see* burial grounds
Center for UFO Studies (CUFOS) 72; *see also* Hynek, J. Allen
CGI (computer graphics) 77, 164
Champ (lake monster) 60–61
charging for services 16, 40, 91, 93, 97, 208
Chasing UFOs (TV show) 80
cherry-picking (evidence) 178, 181
children and ARIGs 95, 99, 140, 219, 229
Chopra, Deepak 137
chupacabra 64, 122, 152, 168
citizen science 17–19, 55, 215
Civilian Saucer Intelligence 70
clairvoyance 30, 48, 135
close encounters (of the third kind) 74, 79
closed-minded 4, 63, 120, 134, 199
Coast to Coast AM 42–43
Cock Lane ghost 45
Coleman, Loren 65, 129
Committee for the Scientific Investigation of Claims of the Paranormal (CSICOP) 73–74
common sense 10, 104, 114, 135
communalism (scientific norm) 103, 112, 210
communication with dead 5, 41, 46, 81, 82–83, 137, 165, 178; *see also* EVP
competition (between ARIGs) 55, 128, 144, 202
conflict (between ARIGs) 55, 96–97, 173, 214; *see also* paranormal unity
cooperation (between ARIGs) 21, 63, 210, 220, 221
community 42, 54, 84, 86, 93, 105, 121, 164, 204, 221; ARIGs 21, 90, 103, 121, 122, 143, 191, 195, 206, 214, 216, 220; cryptozoology 61, 62, 110, 122–123, 196; ghost hunting 51, 60; para-

Index

normal 37, 45, 92, 120, 133, 139, 194, 204, 208; scientific 2, 11, 24, 25, 26, 44, 46, 55, 70, 100, 102, 103, 109, 110, 112, 114, 117, 120, 122, 123, 125, 126, 127, 129, 143, 153, 154, 182, 186, 187, 209, 212; UFO 61, 71, 73, 74, 125, 128
community service 54, 86, 93, 224
Condon Report 71
conferences 14, 61, 65, 73, 76, 79, 88, 89, 116, 174, 231
confidentiality (client) 156
conspiracies 35, 43, 89, 197; UFOs 32, 42, 69, 70, 71, 72, 73, 77
Contact in the Desert 76
contactees: Bigfoot 64; UFO 74–75, 79
contagion (of experiences) 32–33, 219
counter culture 34
cranks 120, 194–195
Creationism (Creationists) 66, 192
critical thinking 89, 93, 94, 183–184, 187, 195, 198–199, 204–205, 207, 209, 212, 214, 215, 218, 220, 229
cryptids origin of term 2, 56, 232; theories about 151–153; UFOs and 89; *see also* Bigfoot; blue dogs; cats; Champ; chupacabra; dinosaurs; dogman; Loch Ness; monster (lake); Mothman; sea serpents; swamp/skunk ape; UFOs; wood ape; Yeren; Yeti; Yowie
cryptozoologist 56, 59, 62–65, 66, 68, 87, 89, 96, 105, 118, 123, 167, 178, 215; professional 65–68, 120
cryptozoology 19, 21, 23, 25, 38, 56–68, 87, 92, 95, 110, 120–121, 127, 128, 129, 130–131, 147, 151–152, 178, 181, 187, 188, 191, 193, 197, 211; academia and 63, 65, 66, 92, 110, 120; skeptics and 62, 66, 122, 197, 199, 206

Dahinden, Rene 60, 98, 108
Darwin, Charles 57, 114
databases 18, 23, 24, 63, 71, 72, 145–146, 155, 216, 233
debunking 7, 46, 50, 59, 165, 231
demarcation problem 188; *see also* Pseudoscience
democratization (of investigation) 39, 119, 121, 122, 123, 208, 209, 210
demons/demonic 1, 22, 29, 30, 41, 51, 85, 99, 124, 127, 135, 147, 151, 171, 174, 175–176, 177, 178, 195, 196, 201, 205, 219, 224, 228
Destination America 40, 175
Did you see that? 200
dimensions (alternative) 64, 65, 125, 127, 137, 148, 150, 152, 165, 176, 196, 207, 219, 225, 232
dinosaurs (living) 66, 151, 232
disinterestedness (scientific norm) 112
Disotell, Todd 65
DNA 59, 65, 84, 118, 120, 157, 167–168, 192; Bigfoot 63, 67, 130–131, 234
do-it-yourself (DIY) 39, 90, 164, 222
dogman 22, 62

Dolby, R.G.A. 11, 113, 126, 127
dowsing 142, 172, 174, 202
drones (aerial) 81, 150
Dyer, Rick 67

e-books 227
ectoplasm 41, 50, 168–169
Edison, Thomas 139
Einstein, Albert 25, 114, 139, 148, 226
electricity 46, 81, 82, 83, 133, 170, 171
electromagnetic fields 49, 135, 148, 172, 203
EMF meters 5, 6, 49, 82, 102, 159, 202, 203
empirical evidence 103, 107, 115, 140, 141, 156, 157, 160
energy 41, 49, 51, 81, 83, 101, 136–139, 147, 148, 163, 165, 169, 171, 174, 177, 190, 198, 203, 221, 224, 225, 226, 227; *see also* electricity; electromagnetic fields
enlightenment 16, 29, 125, 212
environmental factors 17, 47, 49, 106, 140, 141, 144, 148, 156, 166, 172; *see also* space weather; moon
equipment, audio 61, 82, 159, 165; cost 83, 97, 154, 159; ghost hunting 5, 39, 41, 51, 54, 82–84, 141, 142, 202; limitations 81, 141, 156, 160, 162, 219; use 5, 21, 25, 53, 54, 61, 81–84, 86, 97, 133, 140–143, 156, 159, 160, 172, 217, 222, 223, 225, 226; *see also* sciencey
Erickson project 67
ESP 27, 30, 193; *see also* anomalous cognition; psi
ethics 20, 59, 68, 94, 99, 176, 180, 219, 229
evidence: anecdotal *see* anecdotes; eyewitness reports/ testimony; audio 61, 140, 155, 164–167, 182, 202, 203, 219; belief outweighs 10, 30, 53, 94, 96, 111, 125, 156, 160, 182–186, 189, 197, 212, 214, 215, 224; context 102, 144, 156, 184, 213; credibility 8, 63, 103, 129, 130, 140, 153, 158, 158, 190, 192, 195, 213, 215; physical 130, 157, 167–169, 219; physical sensations 169–170; psychic 142, 143, 173–174, 195; subjective 142, 144, 157–159, 160, 167, 169–170, 190, 219; video 32, 34, 59, 60, 77, 84, 151, 152, 156, 162–164, 167, 183, 219
evolution 12, 31, 66, 108, 146, 192, 201, 210
EVP (electronic voice phenomenon) 7, 49, 94, 159, 164–166, 182, 202, 203
exorcism 6, 52, 93, 99, 124, 175, 176, 222
experience (importance of) 6, 10, 16, 17, 25, 27, 33, 37, 87, 97, 103, 157, 159, 171, 210, 208
experiments 48, 49, 92, 102–103, 105–107, 140, 142, 147, 156, 172, 174, 181, 190, 191, 197, 222, 227
expertise (ARIGs) 11, 43, 52, 61, 63, 70, 86, 92, 105, 121, 124, 128, 130, 134, 139, 145, 194, 202, 203, 207, 208, 210, 224, 226, 230
extraordinary claims require extraordinary evidence 111, 113, 121, 123, 213
extraterrestrials 74, 150, 172, 177; *see also* hypothesis, extraterrestrial (ETH)

eyewitness reports/testimony 8, 18, 23, 57, 84, 95, 143, 144, 157–159; problems 7, 9, 23, 48, 63, 66, 75, 107, 134, 157–158, 206
Eyewitness Testimony 158

Facebook 43
Falcon project 67
false positive 182, 215
Fate 35, 79
Finding Bigfoot 33, 40, 62, 63, 67, 134, 205
Flamel College 124
flaps 32, 75
flash mobs 45
flying saucers 34, 69, 70, 76, 79
folklore 17, 57, 58, 62, 66, 73, 74, 92, 93, 95 110, 145, 147, 175, 176, 177, 178
footprints 58, 60, 63, 95, 152, 160, 167, 182, 222
Fort, Charles 21, 35
Fort Mifflin 5, 7, 10, 51, 183
Fortean Times 35
forums (online) 39, 43, 44, 76, 77, 84, 95, 96, 122
French, Christopher 129
Friedman, Stanton 77, 129
fringe subjects 3, 4, 12, 13, 21–22, 25, 28, 30, 31, 34, 43, 47, 62, 73, 93, 96, 110–111, 121, 122, 123, 124, 127, 188, 193–197, 207, 213

Gaddis, Vincent 34
Ganzfeld experiment 135
Gardner, Martin 195
geologic stress 49, 148
geomagnetism 49, 171, 172
Georgia Bigfoot hoax 67
Germany 20, 52
Gettysburg, Pennsylvania 47, 145
Ghost Adventures 20, 53, 134, 229, 231
The Ghost Club 45, 46, 50, 223
Ghost Hunters 16, 19, 20, 38, 39, 40, 45, 52, 53, 54, 62, 81, 134, 201, 202, 203, 229; accusation of faking 204; *see also* TAPS
Ghost Hunting: A Practical Guide 222
ghost hunting guidebooks *see* how-to books
Ghost Hunting 101 228
Ghost Research Society 50
Ghost Science 224–226
ghost science/ghostology 49, 118, 139, 143, 188, 191, 211, 224–226
Ghost Seekers Field Guide 228
ghost tours 14, 16, 38, 51, 89, 92, 93, 221, 224
Ghostbusters 41, 81, 83, 89–90, 91, 92, 169, 222, 225
giant squid 57, 66
Gigantopithecus 63
Gordon, Stan 75
GPS 81
Green Swamp (Florida) 205–206
guidebooks *see* how-to books

Haack, Susan 185
hair (as evidence) 63, 167–168
Hangar 18 77
haunted houses 1, 17, 27, 34, 38, 45, 46, 50, 136, 203
Haunted Media 82
hauntrepreneur 38
Heaven's Gate 77
hero 12, 39, 53, 134
Heuvelmans, Bernard 57, 60, 67
high strangeness 35, 196
Hill, Betty and Barney 74
historical records, obtaining/using 69, 145
History Channel 38, 80
hoaxes/hoaxers 24, 46, 50, 58, 59, 60, 63, 67, 69, 70, 77, 98, 120, 141, 149, 150, 151, 152, 164, 166, 181, 198, 215, 219, 226
hominology 59
Hopkins, Budd 75, 129
Houdini, Harry 46, 213
How to Be a Ghost Hunter 223
how-to books 18, 90, 94–96, 221–230
How to Hunt Ghosts 94, 226–227
Hunting Monsters 123
Huxley, Thomas 57
Hyman's Categorical Imperative 234
Hynek, J. Allen 35, 72, 73, 74
hypnosis (hypnotic regression) 74, 75
hypothesis 105, 106, 136, 141, 146, 180, 184, 214, 215, 219; extraterrestrial (ETH) 32, 95, 150

ideological conviction 180
In Search of... 1, 35, 36, 52, 79, 134
Independent Investigation Group (IIG) 20, 52, 92
International Flying Saucer Bureau (IFSB) 70
International Metaphysical University 124
International UFO Congress 76
Internet 14, 15, 19, 20, 21, 23, 25, 28, 35, 36, 39, 42–44, 51, 55, 62, 65, 76, 77, 79, 84–87, 95, 96, 151, 207; *see also* forums (online)
investigation, process of 3, 16, 37, 41, 62, 85, 86, 91, 106, 143, 144, 145, 165, 180–186, 198, 218, 222, 225, 230
investigation, sham *see* sham inquiry
ion generators 159
Italy 20, 52
ivory tower 63, 104, 121

Jacobs, David 75
Jersey Devil 32, 64

Keel, John 35, 120, 128
Ketchum, Melba 67, 130, 234
Keyhoe, Donald 71, 74
Koestler Parapsychology Unit 48, 129
Korbus, Jason 201–205, 206
Krantz, Grover 60, 63, 65

LARPing 208; *see also* ARIGS, playing pretend scientist
LeBeau Plantation 232
legend-tripping 170
Ley, Willy 56
life after death 25, 45, 46, 48, 51, 84, 94, 125, 128, 137, 138, 143, 191, 201, 222
lights out 162, 233
Loch Ness 1, 34, 61; America's 61
Loxton, Daniel 62, 207, 211
Lyell, Charles 57
Lyons, Sherri 22, 188

Mack, John 75, 129, 194
magnetic fields *see* electromagnetic fields
Mars Curiosity Rover 79
mass hysteria *see* contagion
materialism 125, 139, 191
matrixing *see* pareidolia
Meetup 43
Meldrum, Jeff 65
memory 8, 9, 17, 33, 51, 73, 74, 148, 158, 204, 218
Men in Black 76, 77, 79
Menzel, Donald 73
Merton, Robert 111, 198
meteors/meteorites 75, 79, 189
military (U.S.) 34, 69, 70, 71, 73, 76, 118, 120, 146, 150, 151, 197
Ministry of Defense (U.K.) 70
Minnesota Iceman 60, 67
mist (ghostly) 148, 162, 164, 170, 182
Moneymaker, Matt 63
Monster Hunters (book) 22
Monster Quest 62, 134
monsters, lake 1, 34, 60, 61, 66, 232
moon 49, 144, 150, 171
Most Haunted 38, 52-53
Mothman 22, 32, 62
motivated reasoning 182
MUFON 23, 71-73
Murdie, Alan 223
Myrtles Plantation 145
MySpace 43, 202

Nancy Drew 39
NASA 78, 79
National Enquirer 34
National Investigation Committee of Aerial Phenomenon (NICAP) 71
National Science Foundation 100
National UFO Research Center (NUFORC) 71
natural causes 182, 184, 186
Natural History of Ghosts 1 21
Nature (journal) 46
Nelson, Robert, III 201-205, 206
Nevada Institute of Paranormal Studies 124
New Age (ideas) 34, 71, 80, 125, 137, 172, 176-177, 223

new species 51, 66
night vision *see* cameras, night vision
Nobel prize 110
Nostradamus 2
Not Your Typical Bigfoot Movie 234

objectivity 81, 103, 133, 143, 166, 214
occult 34, 45, 64, 74, 126, 142, 172- 174, 195, 196, 212, 229
occulture 126; *see also* reenchantment
O'Keefe, Ciaran 41
Olympic project 66
Open Minds 77
orang pendek 130, 131
orbs 128, 162-163, 182, 219, 224
ostension 17
The Other Side: A Teen's Guide to Ghost Hunting and the Paranormal 229
Ouija board 172, 173-174, 205, 224, 227, 229
Owen, Richard 57, 59

paradigm shift 195
paranormal: academic study 15, 46-48, 73, 93, 108, 110, 119-131; ARIG objection 31, 206; belief in 2, 3, 10, 11, 27-29, 30, 33, 40, 41-42, 46, 48, 94, 98, 119-126, 177, 182, 186, 197, 205, 206, 207, 210, 212, 214, 215; commodification 38, 84 (*see also* tourism); criticism 6, 11, 30, 33, 36-37, 39, 49, 50, 60, 62, 77, 123, 130, 181 (*see also* skepticism); defined 15, 23, 29, 30, 186; encounters 6, 17, 27, 40, 63, 75, 120, 138, 169, 170-171, 197, 199; fiction 27, 31, 34, 37, 38, 40, 41, 42, 44, 65, 79, 80, 83, 169, 175, 177, 178, 224, 227; hobby 6, 13, 15, 22, 45, 55, 98, 124, 140, 143, 205, 206, 221, 228, 229, 230; hot spots 40, 143, 145, 196 (*see also* high strangeness); investigation 1, 3, 4, 5-7, 15-16, 18, 20, 31, 41, 42, 51, 53, 62, 84, 91, 93, 119, 143, 181, 183, 184, 185, 201-204, 205, 215-220, 221, 230; media 1, 2, 3, 27-28, 29, 32, 34-44, 49, 61-61, 64, 73, 77, 79-80, 135-136, 152, 164, 175, 212, 215; new frontier of science 34, 25-26, 121, 143, 175, 189, 225; normal 44, 207; particularists vs. generalists 29; popularity 2, 3, 19, 23, 27-28, 32, 34, 35, 38-29, 42, 43, 45, 50, 52, 61, 65, 79, 93, 170, 192, 209; scientific rejection 57, 70, 73, 81, 108, 119-123, 153; technology *see* ARIGs, technology; television 3, 16, 19, 27, 31, 32, 34, 35-40, 41, 42, 52-54, 79-80, 90-91, 130, 135-136, 222, 226, 228 (*see also Ghost Adventures*; *Ghost Hunters*; *In Search of*; *Most Haunted*; *Sightings*; *Unsolved Mysteries*); theories 19, 92, 143, 146-153, 193, 197, 210, 211, 225; vs. supernatural 29-31, 119
Paranormal America 3, 17, 28, 207
Paranormal Media 27-28, 34
Paranormal State 53, 54, 55, 134, 135
paranormal unity 55, 63
parapsychology 11, 15, 30, 34, 35, 41, 47-49,

55, 90, 93, 119, 129, 166, 190–191, 193, 196, 201, 222
Parapsychology: A Handbook for the 21st Century 47, 190
pareidolia 148, 152, 159, 161, 166
Paris, Antonio 231
pattern-seeking 9, 197; *see also* pareidolia
Patterson-Gimlin (PG) film 32, 34, 59–60
peer review 21, 101, 113, 127, 129, 133, 140, 194, 216, 220
pendulum 142, 174–175, 227
Pennsylvania State University 53, 55
people as instruments 140, 142, 143, 144, 157–159, 169, 170, 173, 182, 190, 217, 219
perception 29, 54, 79, 110, 136, 139, 144, 157–158, 170, 172, 181, 182, 218; errors 10, 157; *see also* ESP
Persinger, Michael 135, 171
phenomenology 181, 208
Philosophy of Pseudoscience 103
phone apps 149, 151, 164, 224
photographs, evidence *see* evidence, visual; cameras
podcasts 62, 87, 95, 96, 204
poltergeists 34, 46, 48, 171, 177, 186
portals 51, 148, 152, 174, 196, 205; *see also* dimensions, alternative
prediction (scientific) 106, 147, 216
Price, Harry 49–50, 81, 83, 108, 201
primates, anomalous 60, 120, 130, 131; *see also* Bigfoot; *Gigantopithecus*
priming 17, 33, 166, 172, 182
Proceedings of the Royal Academy of Science 131
Project BlueBook 70, 72, 79
Project Grudge 69
Project Saucer 69
Project Sign 69
Project UFO (tv show) 79
proof 16, 29, 46, 60, 61, 84, 98, 111, 132, 138, 154, 156, 159, 168, 169, 183, 186, 187, 190, 205, 215
protection (from entities) 85, 142–143, 172, 175, 179
Prothero, Donald 62, 207, 211, 215
provoking spirits 174, 202
pseudoscience 2, 41, 93, 104, 115, 117, 187–195, 210; characteristics 187–190; in media 191–192
psi 4 9, 73, 191, 197, 202; *see also* ESP
psychical research 45–47, 48, 49, 50, 82, 90, 119, 127, 128; *see also* parapsychology; Society of Psychical Research
psychics/mediums 28, 30, 34, 45, 46, 48, 50, 74, 95, 129, 135, 142, 168, 169, 173, 202, 213, 225
psychology 3, 9, 24, 49, 51, 73, 74, 75, 90, 97, 109, 110, 133, 134, 139, 147, 190, 194, 223; anomalistic 48, 129
public definition of: 104–105; ARIGs and 6, 7, 11–12, 20, 22, 25, 38, 39, 53, 54, 73, 86, 87, 92–94, 118, 121, 156, 186, 199, 208, 209, 210, 212, 216; paranormal ideas 3, 14, 27–29, 31, 34, 35, 37, 41–41, 45, 57, 58, 61, 67, 70, 71, 79–80, 90, 123–124, 208, 221
public understanding of science 1, 2, 13, 41, 81, 100–102, 109–110, 114–115, 126–127, 189, 191–192, 208, 209, 210, 211–212

quantum (effects/theory) 49, 118, 133, 136, 137, 138, 139, 143, 148, 195; *see also* sciencey
quantum flapdoodle 137

Radford, Benjamin 61, 62, 122, 139, 204, 213, 217, 230
radio 34, 43, 50, 77, 79, 81–83, 87, 95, 134, 166, 172
Raelians 77, 176
Randi, James 204, 213
rational inquiry 104, 183, 184, 215–216, 230
Raudive voices *see* EVP
Real Science 102
reality television 31, 36–40, 42, 53, 62, 80, 90, 91, 163, 173, 206
red panda effect 32
reenchantment 17, 23, 176
religion 24, 27, 97, 116, 125–126, 135, 156, 175–177; *see also* Bible; Bigfoot, spirit; Creationism; demons/demonic; Heaven's Gate; Raelians; Spiritualism; UFOs and religion
reveal (evidence) 54, 134, 203
Rhine, J.B. 48
Rhine, Louisa 48
Rhine Research Center 124
Robertson panel 70
rock-throwing 171
rods (sky fish) 150, 163–4
Roll, William 201, 204
Rose Hall, Jamaica 231
Roswell 77

Sagan, Carl 73, 120, 211, 213, 226
Sanderson, Ivan 34, 57, 58, 60, 67, 108
Sasquatch 58; *see also* Bigfoot
Satanic Panic 75
science: alternatives 107, 172, 188 (*see also* pseudoscience; religion); ARIGs misuse of 137, 139, 146–147, 153–155, 160, 182, 187, 225, 226 (*see also* ARIGs, playing pretend scientist; sciencey; scientific); as authority 2, 100–101, 102, 114, 132–133, 209, 221; boundaries 8, 22, 30, 109, 116, 125, 187, 188, 191; change rules 180, 186, 196; as collective enterprise 2, 100, 102, 103–4, 109, 112, 117, 209 (*see also* community, scientific; communalism); criticism 49, 101, 103, 112–113, 199, 216 (*see also* peer review); defined 8, 102–104, 107, 108–109, 112–113, 116–117, 186, 209; deviant 193, 194; doesn't know everything 108; fringe 189, 193–194, 197; honorific use 92,

Index

109, 115; imitation 37, 38, 100, 115, 117, 132–133, 21 (*see also* ARIGs, playing pretend scientist; scientifical); jargon 74, 102, 116, 118, 136; as method *see* scientific method; misunderstanding 101, 133, 136–142, 143, 186, 192, 212, 226; and paranormal 1, 3, 7, 13, 18, 25, 26, 30, 86, 108, 110, 119–131, 187, 203, 225 (*see also* parapsychology; sciencey; scientifical); portrayal 40–41, 81, 113, 114, 117, 121, 132, 133 134–135, 155, 209 (*see also* science; scientificity); as privileged 103, 107–108, 117; and public 100–101, 104, 114, 115, 123, 209, 212, 221 (*see also* public understanding of science); reliability 8, 102, 104, 107–108, 109, 115, 209; social aspects 106–107 (*see also* community, scientific); and technology 81, 83, 102 (*see also* ARIGs, use of technology; sciencey)
Science (journal) 46, 47
science appreciation 3, 18, 100, 211–212
Science in American Society 110
science literacy 100, 101, 115, 134, 208, 209, 211, 221, 226; *see also* public understanding of science
Science Talk 101
science talk 100, 101–102, 115, 116, 118
sciencey 1, 40, 84, 100, 116–118, 133, 134, 136–140, 148, 153, 178, 187, 189, 196, 204, 210, 224, 225, 226, 228, 234
scientese 115
scientific ethos 109, 111–113, 116, 123, 186, 189–190
scientific expertise *see* expertise
scientific method/process 8, 12, 101, 103, 104–108, 111, 140–144, 153–154, 157
Scientific Paranormal Investigation 217, 230
scientific theories *see* theory, definition of
scientifical 2, 22, 118, 134, 220
scientificity 24–25, 153–155, 156, 172, 228, 210
scientist, as professional 13, 22, 25, 57, 65, 109, 114, 116–117, 120, 121, 123, 125, 134; stereotype 104, 114, 115
Scooby Doo 33, 39
sea serpents 34, 56, 57, 59, 84, 188
Searching for Sasquatch 21
self-publishing 128, 221
Selling Science 132
serious leisure 14–15, 22, 54, 97, 108, 208, 215
shadow people 6, 36, 79, 148
sham inquiry 2, 185, 212–214, 215, 217, 220
Shermer, Michael 204
Shuker, Karl 65, 129
Sightings 35, 38, 201
Six Million Dollar Man 62
skepticism 6–9, 10, 15, 33, 73, 84, 112, 133, 158–159, 182, 187, 189, 190, 193, 197–199, 206, 229
sleep paralysis 205
Slenderman 44
Slick, Tom 58
smudging 177–178

social media (networks) 35, 39, 43, 77, 231
Society of Psychical Research (SPR) 45, 46, 47, 50, 91, 92, 124; American 45, 46, 47, 50
software 43, 77, 159, 164
solar activity 49, 148, 171
sounds, anomalous 61, 165, 166–167; *see also* EVP
space brothers 74, 125
space weather 49, 171
special pleading 190
Species, Serpents, Spirits and Skulls 22
spirit cleansing 178–179
spiritualism 22, 28, 82, 176, 188
squatchin' 62, 134
Standing, Todd 67
statistics, use of 48, 49, 73, 93, 128, 133, 138, 146, 167, 215
The Stone Tape 41
stone/water tape 49, 137, 147–148
Strange Frequencies Radio 204
Strieber, Whitley 75
supernatural agents 30, 119, 197
supernatural creep 64, 195–196, 197
supernatural, definition of 29–30
supernatural vs. paranormal *see* paranormal vs. supernatural
Survivorman: Bigfoot 67
swamp/skunk ape 58–59, 205–206
Sykes, Bryan 65–66, 110, 118, 120, 130–131, 206

TAPS (The Atlantic Paranormal Society) 16, 20, 38, 45, 53, 54, 202, 229; family 20–21
technology *see* ARIGs, use of technology
telegraph 81, 82
telekinesis/psychokenesis 30, 172
telepathy 46, 48, 222; *see also* ESP
temperature measurements 82, 141, 159
theory, definition of 146
theory, surrogates for 146–147
thought forms 65, 148, 151
Thurs, Daniel 101, 115, 192
time travel 148, 150, 196
tourism 14, 16, 28, 38, 51, 61, 77, 89, 92, 93, 221, 224; *see also* ghost tours
Tracking the Chupacabra 122
tri-field meter 202; *see also* EMF meters
tribalism 214
trigger object 142, 174; *see also* provoking spirits
tulpas *see* thought forms

UFO Contact Center International 75
ufology 22, 25, 35, 69–80, 95, 97, 120, 127, 128, 147, 181, 187, 188, 191, 193, 197, 211
UFOs: crashes 77–79; methods of investigation 24, 71, 72; nuts and bolts 76, 77; online 35, 42, 44, 76, 77, 79, 96, 151; as paranormal 23, 29, 31, 73; religion and 74, 125, 176, 177, 234; scientific examination of 14, 57, 70, 71,

73, 123, 128, 192; TV and film 32, 35, 38, 41, 76, 79–80
Ultimate Ghost Hunters Field Guide 228
Ultimate Ghost Hunting Guide 227
Ultimate Ghost Tech 139, 224
ultraterrestrials 150; *see also* dimensions (alternative)
unfalsifiable 189, 191, 195
United Kingdom (U.K.) 19, 38, 50, 70, 76, 172
universalism (scientific norm) 112
Unsolved Mysteries 35, 36, 38, 52, 79, 134, 201

Vallee, Jacques 73–74
Velikovsky, Immanuel 34, 126–128, 194
video game controllers 81
von Daniken, Erich 80

Walton, Travis 74
Warren, Joshua 94, 226–227
white lab coats 104, 114, 118
White Noise (film) 41
white noise generators 159

wild men/ hairy men 32, 58, 59, 60, 75, 79, 178
Wilson, Vince 139, 224–226
Windbridge Institute 48, 129
Wiseman, Richard 129, 172
wood ape 64, 161; *see also* Bigfoot
wood knocking 63
Worlds in Collision 126, 128
worldview 9–11, 28, 42, 95, 136, 157, 185, 195; paranormal 27, 98, 183, 206, 207, 213, 220
Worldviews 9

The X Files 32, 35, 37, 79

Yeren 59
Yeti 34, 58, 59, 61, 65, 120, 130
YouTube 19, 76, 87, 206
Yowie 59

zooform phenomena 127, 151, 196
zoology 31, 110, 130, 131, 133, 234; historical 56–59; romantic 57